Translational Medicine – What, Why and How: An International Perspective

Translational Research in Biomedicine

Vol. 3

Series Editor

Samuel H.H. Chan Kaohsiung

Associate Editor

Julie Y.H. Chan Kaohsiung

Translational Medicine – What, Why and How: An International Perspective

Volume Editors

Barbara Alving Bethesda, Md.
Kerong Dai Shanghai
Samuel H.H. Chan Kaohsiung

14 figures, 7 in color, and 14 tables, 2013

Basel · Freiburg · Paris · London · New York · New Delhi · Bangkok · Beijing · Tokyo · Kuala Lumpur · Singapore · Sydney

Barbara Alving, MD
Uniformed Services
University of the Health Sciences
7118 Glenbrook Road
Bethesda, MD 20814 (USA)

Kerong Dai, MD
Clinical Translational Center of Stem Cell and
Regenerative Medicine
Ninth People's Hospital
Shanghai Jiao Tong University
School of Medicine
Shanghai (China)

Samuel H.H. Chan, PhD
Center for Translational Research in Biomedical
Sciences
Kaohsiung Chang Gung Memorial Hospital
123, Taipei Road
Kaohsiung 83801(Taiwan)

The Chang Gung Medical Foundation is the patron of this book series.

Library of Congress Cataloging-in-Publication Data

Translational medicine : what, why, and how : an international perspective /
volume editors, Barbara Alving, Kerong Dai, Samuel H.H. Chan.
p. ; cm. -- (Translational research in biomedicine, ISSN 1662-405X ;
v. 3)
Includes bibliographical references and indexes.
ISBN 978-3-318-02284-1 (hard cover : alk. paper) -- ISBN 978-3-318-02285-8
(electronic version)
I. Alving, Barbara M. II. Dai, Kerong. III. Chan, Samuel H. H. IV. Series:
Translational research in biomedicine ; v. 3. 1662-405X
[DNLM: 1. Translational Medical Research. 2. Academic Medical
Centers--trends. W 20.55.T7]

610.72--dc23

2012041085

www.karger.com
Printed in Switzerland on acid-free and non-aging paper (ISO 9706) by Reinhardt Druck, Basel
ISSN 1662–405X
e-ISSN 1662–4068
ISBN 978–3–318–02284–1
e-ISBN 978–3–318–02285–8

Contents

Foreword

Welcome to volume 3 of *Translational Research in Biomedicine,* a monograph series dedicated to the dissemination of seminal information in contemporary biomedicine with a translational orientation. This volume marks a new milestone of this series. Beginning with this volume, *Translational Research in Biomedicine* will be published under the generous patronage of Chang Gung Medical Foundation, Taiwan. This patronage will reduce substantially the increasing financial constraints on scientific publication, and allow us to concentrate on publishing timely and crucial themes in translational medicine.

This volume is designed to be a key reference for translational medicine. During the last decade, translational medicine has become a household word in the arena of medical research and beyond. Closer scrutiny, however, has revealed that the term 'translational medicine' has been interpreted and presented widely, ranging from pure rhetoric in political orations, golden geese to entrepreneurs or shareholders, to conscientious solutions to health issues. My personal view is that this multiplicity arises from the lack of a comprehensive overview on the 'what', 'why', and 'how' aspects of translational medicine. This volume is designed to fill this void based on an international perspective.

I wish to express my deepest appreciation to Drs. Barbara Alving and Kerong Dai, whose patience and sustaining efforts have made this timely volume, *Translational Medicine – What, Why and How: An International Perspective,* a reality. Last but not least, the publication of *Translational Research in Biomedicine* would not have been possible without the foresight, enthusiasm, and whole-hearted support of my dear friend, Dr. Thomas Karger.

Samuel H.H. Chan, Kaohsiung
Series Editor

Preface

Although translational medicine is now a common term throughout the scientific world, a comprehensive monograph that addresses the 'what', 'why', and 'how' aspects of translational medicine, particularly from an international perspective, has not been forthcoming. This volume will fill this void. Fourteen chapters written by eminent experts with diversified experience will address three general themes: 'What is translational medicine?', 'Why do governments, research institutions, and other organizations invest in translational medicine?', and 'How is translational medicine carried out internationally?'. In so doing, it is our intention to make this volume a key reference of translational medicine.

Our sincere thanks go to the authors who have kindly consented to contribute their expert opinion to this volume and to Chang Gung Medical Foundation in Taiwan for sponsoring the *Translational Research in Biomedicine* series.

Barbara Alving, Bethesda, Md.
Kerong Dai, Shanghai
Samuel H.H. Chan, Kaohsiung
Volume Editors

Alving B, Dai K, Chan SHH (eds): Translational Medicine – What, Why and How: An International Perspective.
Transl Res Biomed. Basel, Karger, 2013, vol 3, pp 1–5 (DOI: 10.1159/000343146)

The What and Why of Translational Medicine

Barbara Alving[a] · Kerong Dai[b] · Samuel H.H. Chan[c]

[a]Uniformed Services, University of the Health Sciences, Bethesda, Md., USA; [b]Clinical Translational Center of Stem Cell and Regenerative Medicine, Shanghai Jiao Tong University School of Medicine, Shanghai, PR China; [c]Center for Translational Research in Biomedical Sciences, Chang Gung Medical Center, Kaohsiung, Taiwan

Abstract

During the past six years, multiple countries have embraced the concept of translational medicine. This is generally understood as improving prevention and treatment strategies for patients and for communities through translating discoveries in basic and clinical research into products and practices in an efficient, cost-effective manner. Although the emphasis on the different domains of translation vary among countries and institutions, all are developing an increasing ability to share ideas and practices through an international language of translation.

Translation, or the transfer of knowledge, scientific discovery, and practice from one 'domain' into the other has always been a part of the research process, although the activity is recognized and practiced to greater or lesser degrees as a function of the discipline. For example, translation into products is an inherent part of training in engineering and needs little or no discussion. During the past 60 years, the majority of biomedical researchers in the USA have focused on basic science, which has been largely encouraged and funded by the National Institutes of Health (NIH). Investigators are promoted at academic health centers (AHCs) based primarily on their publication record and on their ability to garner continued NIH funding. They have assumed that the fruits of their research would be used by those in other disciplines or organizations, such as industry, who found their work of interest. The US Congress, which authorizes funding for the NIH, has always had an interest in translation and has mandated that 2.5% of the NIH budget be used for Small Business Innovation Research. The passage of the Bayh-Dole Act in 1980, which provided for a university, small business, or non-profit institution to have ownership of an invention arising from federally funded research and required the institution to share any royalties with the inventors, incentivized any number of investigators and their institutions to work in public-private partnerships with now well-defined codes of conduct. However, the AHCs in the USA were not encouraged or funded to emphasize translational activities until the NIH, under the directorship of Elias Zerhouni, announced funding for a large program known as the Clinical and Translational Science Awards (CTSAs). Between 2006 and 2011, a total of 60 AHCs in the USA received such awards, which are 5 years in duration, followed by recompetition [1]. The awardee AHCs are developing infrastructure to translate basic laboratory discoveries into treatment and prevention strategies in a more efficient and cost-effective manner. In the Chinese mainland, more than 34 centers for translational research were developed between 2006 and early 2011, and a total of 51 centers were established by the end of 2011. In these centers, there is strong interdisciplinary research that tends to be focused on the areas of expertise of the given centers [2, 3].

What Is Translational Medicine?

Translational medicine, while identified in the USA with a very large highly funded NIH program, is really a philosophy and a culture. It can best be considered as efforts to move the results of basic research forward into preclinical and clinical research and into the community in a cost-effective and efficient manner. The outcome will be to impact health through improved prevention practices and development of novel drugs and devices. In a broader sense, translation medicine can be taken as a bench to bedside and back approach to foster communication between the preclinical and clinical communities [4] in the formulation of our modern understanding of health provision [5]. Translational medicine activities are complex and extend beyond AHCs, involving partnerships with private industry, businesses, investors, and patient advocacy groups, as well as with government agencies that fund or regulate translational research [6–8]. A key aspect of this initiative is to train investigators to conduct translational research as interdisciplinary teams.

Trochim et al. [9] have recently published a review of the different ways in which the stages or phases of translation (T) have been defined during the past decade. Most authors identify the first phase of translation (T1) as making the connection between laboratory research and early phase trials in human subjects. The expansion of early phase studies into larger clinical trials has been described as T2; studies of clinical effectiveness and translation into practice have been described as T3 with T4, representing the closing of the loop and translating back into the laboratory, identifying further studies needed or alternatively, conducting 'outcomes research' [7–9]. However, such stages tend to be artificial and do not really reflect the iterative, complex nature of basic and clinical research and incorporation into clinical practice. The most compelling reasons to define the stages of clinical and translational research are for evaluation and resource allocation. For such purposes, rather than a broad multiphased approach, it is preferable to use a more detailed 'process marker' model [9] that identifies specific operationally definable markers along the translational continuum, making it possible to conduct process measurement and monitoring. The use of process marker models for evaluation will be discussed in this volume in the chapter by Kane, Rubio, and Trochim [pp. 110–119].

Why Do Governments, Research Institutions and Other Organizations Invest in Translational Medicine?

Nations throughout the world recognize that the processes of moving discoveries in basic laboratory research into new treatments and prevention strategies for improving health can be done in a much more efficient, cost-effective manner.

In this era of increasingly complex science, interdisciplinary teams are required; they may include clinical and basic researchers, informatics experts, bioengineers, biostatisticians, and experts in public health, ethics, law, industry and marketing. Biomedical researchers in translational research centers are being trained in the components of successful drug, device, or biomarker development and on how to recognize the potential commercial value of their work with respect to developing drugs or biomarkers for particular diseases. In many of these centers, technology transfer offices provide components of training, establish policies for partnerships, and foster an environment that promotes public-private collaboration. To increase efficiency, CTSA institutions in the USA have partnered with their schools of business and are learning how to create business plans in their research and how to work more efficiently. In the Chinese mainland, there are a variety of clinical and translational training programs that are committed to implementing the philosophy of translational medicine, introducing overseas experience, and guiding the construction of a transformational medical platform. To provide a broad understanding of training activities around the globe, this volume will highlight training and workforce development in the UK, Taiwan, the Chinese mainland, and the USA.

The obligations of translational medicine centers are to provide service, coordinate the interests of the project teams, and construct dynamic infrastructures. Each center also provides a 'home' where investigators can find help with their clinical research needs, such as consultation in biostatistics and protocol design, as well as preparation for documents to be reviewed by their institutional review boards. The translational medicine centers require sustained funding for the resources that investigators need to design and conduct high-quality clinical research. One way in which to achieve this is to have well-constructed cost-recovery processes [10]. Resources at most AHCs in the USA were not integrated prior to the CTSA initiative. Many are now well-organized and available to clinical investigators from all specialties, and investigators are learning how to use them through formal training. The informatics tools that are being developed for patient recruitment can be shared across all institutions in the USA as well as in other countries.

Policies that are common to universities in a given country and that can be shared and agreed upon internationally would be a great stimulus to international collaborative efforts, especially since most of the large pharmaceutical companies fund clinical trials that are multinational. NIH-sponsored clinical trials are often multinational in scope to assure accrual of patients and diversity in population. On a regulatory level, collaboration and training are underway among the US FDA and the SFDA in China, as well as in India and other countries.

How Is Translational Medicine Carried Out Internationally?

The chapters in this volume are dedicated to the 'how' of translational medicine. How does a country (institution) define the goals of translational medicine efforts and how does an institution or funder evaluate such efforts and determine their merit relative to the financial investment? How does an institution ensure that basic scientists and clinical investigators work together in a productive manner? Since few clinicians have the time or personnel support to be engaged in research, how are institutions finding protected time and ensuring that PhD scientists, nurses, and other personnel receive the training to become much more engaged in clinical research, thus relieving some of the time constraints of the practicing clinicians? In the USA, this shortage of clinical investigators is beginning to be addressed with the very successful Howard Hughes Medical Institute-funded 'Med-Into-Grad' program, in which nonphysicians have the opportunity to take medical courses and to see patients with clinicians to learn in a much more direct way the clinical problems that need to be solved.

A strong philosophy of translation at an institutional or national level provides the permission and impetus for investigators to continue searching and exploring outside their traditional domains and disciplines. Innovation is greatly facilitated by the presence of a solid infrastructure that promotes training, supports intellectual property development, encourages interdisciplinary collaboration in public-private partnerships, and also provides investigators with informatics tools to streamline their clinical research activities. New opportunities in informatics make this decade the right time for collaboration and connection, which leads to translation.

A critical feature of translational philosophy and effort is that it does not have an 'end point' or a finish line. Thus, a translational center is similar to a business venture, which constantly encourages innovation, whether it is in new products or treatment strategies, or in new ways to deliver care to patients and communities to address their greatest needs. In a recent article in the *Harvard Business Review* entitled 'Managing your innovation portfolio', the authors emphasize that in a firm, 70% of the activity is for the routine core activities, 20% to less sure areas, and 10% to transformational high-risk ventures [11]. The transformational initiatives are essential to the health of the firm and funding should come from outside budget cycles. For AHCs, such funding may need to come from the institutions or from other sources, rather than from the government. The authors also state that pipeline development should come from nurturing a few promising ideas with 'nonfinancial achievement' recognized in the early phases of the project.

Ideally, the multinational emphasis on translation medicine will lead to the formation of efficient networks of researchers focused on their areas of expertise and supported by common policies among nations, as well as by integrated informatics systems. They will be supported at their own institutions by clinical research resources, both personnel and facilities, thus increasing the efficiency and quality of research and its benefits to patients and communities.

References

1 Zerhouni EA: Translational and clinical science – time for a new vision. N Engl J Med 2005;353:1621–1623.
2 Dai K: The development strategy of translational research based on China's current national situation; in Sanders S (ed): Selected Presentations from the 2011 Sino-American Symposium on Clinical and Translational Medicine. Washington, Science AAAS, 2011, p 13.
3 Dai K, Yang F: Reflections on building translational medicine platform; in Dai K (ed): Translational Medicine: The Concept, Strategy and Practice, ed 1. Xi'an, Fourth Military Medical University Press, 2012, pp 183–189.
4 Hörig H, Pullman W: From bench to clinic and back: perspective on the 1st IQPC Translational Research Conference. J Transl Med 2004;2:44–51.
5 Sonntag KC: Implementations of translational medicine. J Transl Med 2005;3:33–35.
6 Portilla L, Alving B: Reaping the benefits of biomedical research: partnerships required. Sci Transl Med 2010;2:1–3.
7 Shekhar A, Denne S, Tierney W, Wilkes D, Brater DC: A model for engaging public-private partnerships. Clin Transl Sci 2011;4:80–83.
8 Collins FS: Reengineering translational science: the time is right. Sci Transl Med 2011;3:1–6.
9 Trochim W, Kane C, Graham MJ, Pincus HA: Evaluating translational research: a process marker model. Clin Transl Sci 2011;4:153–162.
10 McCammon MG, Fogg TT, Jacobsen L, Roache J, Sampson R, Bower CL: From free to free market: cost recovery in federally funded clinical research. Sci Transl Med 2012;4:1–5.
11 Nagji B, Tuff G: Managing your innovation portfolio. Harvard Business Rev, May 2012, pp 66–74.

Barbara Alving, MD
Uniformed Services, University of the Health Sciences
7118 Glenbrook Road
Bethesda, MD 20814 (USA)
E-Mail: balving@alving.org

Alving B, Dai K, Chan SHH (eds): Translational Medicine – What, Why and How: An International Perspective.
Transl Res Biomed. Basel, Karger, 2013, vol 3, pp 6–17 (DOI: 10.1159/000343013)

Partnership Model for Academic Health Science Systems to Address the Continuum from Discovery to Care, at Scale

David Fish · Cyril Chantler · Ajay K. Kakkar · Richard Trembath ·
John Tooke · on behalf of UCLPartners

UCLPartners Academic Health Science Partnership, London, UK

Abstract

Academic Medical Centres (AMCs) translate laboratory research into clinical success. University College London Partnership (UCLP) is further designated to translate proven interventions into population health gain and wealth creation. UCLP has grown since 2009 from a partnership of 5 institutions to include 9 universities, 16 major hospitals, community and mental health organisations with a combined annual budget of USD15bn and health care employees of 70,000, serving a local population of 6 million. UCLP is a 'not for profit' company serving the partnership and acting an enabling catalyst. The culture of the partnership is crucial. Progress relies on the shared vision, trust and transparency, rather than further contracts between autonomous partner organisations. This has enabled tangible benefits in major areas of disease burden such as stroke and cancer. UCLP is also responsible for much post-graduate clinical education and leadership development. This creates the workforce that the patients and organisations will need for the anticipated changes to healthcare delivery over te next decade, and serves to bind the partnership to the shared vision.

All nations face the challenge of providing high-quality and cost-effective healthcare that meets the needs of their populations. Demographic shifts with the attendant rise in chronic illness and degenerative processes, coupled with the costs associated with burgeoning technological capacity, increase the difficulties of rising to this challenge, especially given the current global economic outlook. Despite advances stemming from a profound understanding of the biological basis of disease, there is a growing recognition that in order to best address the burden of chronic disease and multiple morbidity, such advances need to be assimilated with appreciation of lifestyle and social determinants of health.

In 2009, the Department of Health in England designated academic health science center (AHSC) status on five academic health alliances after a rigorous vetting process including international peer review of the quality, breadth and scale of medical science, and alignment across the health and academic activities conducted by the centers. Coincident with this, a seminal article by Dzau et al. [1] commented on the misdirected nature of much health research and development. A focus on late-stage disease – the traditional emphasis of the AHSC of the last century – fails to adequately address the challenges of an ageing population, the critical importance of preventive health to mitigate the burden of chronic disease, the potential of personalized or predictive healthcare, and escalating health costs and disparities in both care access and outcomes. In many countries, Dzau et al. argued, such problems are compounded by inadequate continuity of care across community, primary, secondary, and tertiary care settings. Academic health science 'systems', the authors asserted, are ideally poised to become the system integrators, leading the necessary transformation of medicine through the development of a 'discovery-care continuum' capable of closing translational gaps and delivering evidence into practice at a population level.

This vision attunes with the aims of the National Institute of Health Research in the UK to speed the progress of discovery into practice. In England, the National Institute of Health Research has invested GBP 800 million in biomedical research centers and units focused on experimental medicine to close the gap between laboratory research and proof of concept in human studies, and to develop strong links with industry. Such resources are concentrated in the five AHSCs. However, there remain the challenges of delivering new treatments into mainstream clinical practice and to encourage their widespread adoption for optimum use within populations and communities. Striking variations exist in access to, uptake of, and outcomes resulting from medical care, with the dissemination of proven interventions at a rate of circa 3% per annum (it takes on average 17 years from proven value to widespread uptake in clinical practice) [2]. The reliability of the care that is delivered is substantially lower than tolerated in other industries. The biomedical model of healthcare is insufficient to explain and address these gaps.

It is widely acknowledged that there is no single, ideal organizational model for academic health science systems that will deliver the agenda outlined by Dzau et al. [1]. In a recent article, Daniels and Carson 2011 [3] describe two models: itemized contractual relationships or 'the single firm' (a single university and hospital with common leadership). After 3 years of experience, we describe here the evolution of a third approach, based on a formal partnership that seeks to embrace all participants across the health system whilst retaining individual autonomy.

Partnership Model

University College London Partners (UCLPartners) was formed in 2009, initially as a partnership between one university and four teaching hospitals (two specialist hospitals and two full-range hospitals, each with major research and education activities), and formalized through a legal entity – a 'not-for-profit' company limited by guarantee (there are no shareholders). Since then, as a consequence of growing alignment across our system and evidence of both cultural change and delivery, the partnership has evolved to include virtually all major acute/specialist hospitals, mental health services, community service providers and seven universities (combined revenue income is USD 15 billion per year). The partners who employ >70,000 healthcare staff, cover a contiguous population base of more than 5 million across a 40% segment of London, and adjacent counties (see http://www.uclpartners.com for details).

The large-scale system approach adopted by UCLPartners provides the foundation to enhance population health and drive healthcare improvement defined as outcomes that matter to patients per pound spent [4]. The integrated population coverage also fulfills an essential requirement for research activity: the potential to access well-phenotyped and/or genotyped patients and cohorts who can engage in such research, as well as amplifying the impact of subsequent developments. Such scale is, for example, essential in the study of both rare diseases, from which discoveries of generic significance increasingly derive, and the heterogeneous nature of common disorders.

Given the well-recognized problems of converting evidence into clinical practice [2, 5], at the outset we established a learning program to develop and test the system attributes associated with faster delivery into practice of new treatments and innovations, continually applying, adapting, and refining this across a portfolio of over 50 major projects supported by UCLPartners (http://www.uclpartners.com/our-work/). The programs and their leaders were selected through a competitive process that included a review of the proposed projects in terms of the potential to improve both treatment and the health of the populations we serve. The management of the programs encompasses four key components: provider autonomy, alignment, collaboration, and 'exit strategies' (to embed and release new innovations rather than control their routine use centrally).

Provider Autonomy

UCLPartners is a partnership of autonomous healthcare providers and higher education establishments – as opposed to a single entity or complex series of contractual arrangements – based on the shared vision of delivering value in healthcare and with a joint focus on wealth creation. This approach facilitates rapid mobilization of effort and resources to deliver transformational change through clarity of mission, peer pressure, and aligning resources across the health and academic systems. Furthermore, maintaining autonomy, as is well known from many successful 'clusters' of interconnected organizations within one industry (e.g. biotech in Boston), drives innovation and performance through simultaneous collaboration and competition [6]. For example, our partners are collaborating to create an integrated cancer system designed around patient and population need, yet they will continue to compete for market share within that system.

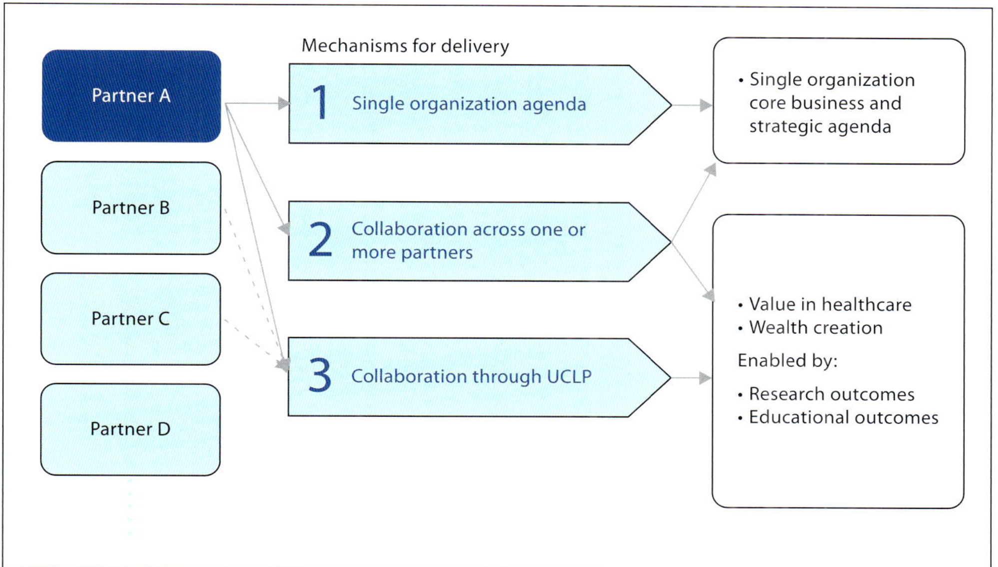

Fig. 1. For each of the partners (e.g. Partner A) there is a perspective of three delivery mechanisms. Most of their resource and attention remains on delivering their core business, and all partners remain autonomous, accountable to their own boards. Collaboration with others can facilitate their own agendas and at times deliver broader benefits for patients and populations. The core focus of UCLPartners is on providing mechanisms that will collectively support delivery of patient and population health gain and the emerging focus to enhance wealth creation. Sharing of information between the partners across traditional boundaries is an increasingly evident enabler.

The partnership approach, therefore, does not replace existing pathways for delivery of individual organizations acting alone or in collaboration with another partner directly where that route is effective. Instead it utilizes the power of the partnership acting together to provide a more appropriate conduit for projects needing greater capacity or system alignment at greater scale than traditionally achieved (fig. 1).

Alignment

Accelerated delivery has been achieved when all parts of the system – patient networks and organizations, our providers and universities, commissioners and government – have aligned in the relentless pursuit of a common goal. This has enabled us to more rapidly mobilize resources and ensured alignment of incentives across the system to facilitate improvement. Recognizing the imperatives of education and training to bind the current and future workforce collectively to the aims of patient and population health gain, and staff who understand approaches to improvement beyond traditional biomedical skills, UCLPartners is increasingly undertaking a lead role in the organization of multiprofessional learning with innovative programs aligned to our purpose. For example, UCLPartners is a designated lead provider for medical and dental education across our partnership (i.e. contractually responsible for junior doctor training), creating modular master's programs inclusive of e-health, leadership and improvement science that will be available to all trainees. A similar program for postgraduate nursing and midwifery development has started recently. This responsibility will drive a new cohort of the workforce aligned to the challenges of the next decade. The alignment with our programs is evident from the planned whole life-course approach to women's health training and education, developed by the program director, and fed into national policy (see Box 1) [7].

Box 1. Women's Health Programme – a whole life-course approach

The Women's Health Programme has actively engaged users – local population, patient focus groups and social networks – to enable from the outset enhanced 'pull through' into practice, and based its subsequent proposals on a life-course model. This offers a more unified and woman-centered approach to health promotion, disease prevention, and management, recognizing predictable events (e.g. pregnancy and childbirth) that have potentially profound effects on women's health in later life and on the next generation, and acknowledging a wide range of other influences on health – from social determinants of health to epigenetics and the biology of ageing.

The life-course model has implications for clinical practice (services becoming more aligned to predictable healthcare needs), research (including individual data linkage studies across generations), education (about healthy lifestyles and personalized medicine), and training (raising professional awareness of the life-course model and its implications). It has formed the basis to develop the systematic alignment of postgraduate training with a women's health life-course rather than traditional departmental/uni-professional or event-related care [7].

Leadership development is an essential enabler for alignment around our vision. Developing multidisciplinary leaders of character and competence is critical. In a unique development, working with the exceptional expertise of the armed services in this area, we have established a UCLPartners Staff College (hosted by one of the partners) to serve as a common source of joint learning across all system partners and professions (including primary care) [8]. The training is reality-based rather than role-playing of contrived scenarios. More than 200 delegates attended the course during the past year. Feedback obtained from the delegates at interview demonstrates an appreciation of the value of this approach by >95% of attendees.

The second key enabler for alignment is shared informatics – focused on the hand-offs and interfaces rather than the traditional single 'ring-fenced' or separate organizational structures. The partnership has appointed a single, chief clinical information officer to support the program of integrating informatics system-wide, joining clinical data across the spectrum of care within a clear governance framework. We will work towards enhanced communication between systems, promoting local innovation rather than a single uniform hardwired approach.

Thirdly, we have dedicated considerable energy to building relationships with and informing the priorities of local commissioners and national bodies to facilitate alignment with our patient-led and population-focused priorities. The size of our partnership, its independence, track record of delivery and academic activity affords us a degree of credibility to provide a 'safe space' for otherwise difficult conversations. This is particularly important in engaging with the primary care structure, which, in the UK, represents the gateway into specialist care for the whole population. In delivering new models of care for stroke and cancer, for example, our commissioners set the parameters for change incorporating challenging timeframes. In response, we were able to rapidly organize ourselves as a partnership to generate and implement locally relevant and system-wide solutions. An example is the transformation of care for hyperacute stroke (see Box 2).

Box 2. Hyper acute stroke unit

In the setting of NHS London's plans to deliver a new model of care for stroke, UCLPartners rapidly facilitated a unified service across our original population base (i.e. North Central London) – a single-point-of-entry hyperacute stroke unit to which all ambulances bring patients with suspected acute stroke, 24/7. UCLPartners extended the model to include a rotational system whereby all stroke physicians share the on-call duties at the hyperacute stroke unit and deliver care across the population at one of the four stroke units, providing early care and rehabilitation closer to home, with common metrics spanning the whole patient pathway [9]. UCLPartners has also led the health-economic evaluation of the pan-London changes: early results show a thrombolysis treatment rate of circa 14% across the capital (more than any other major city internationally), a reduction in mortality of 30% (double the national average improvement), and significant cost reduction reflecting reduced morbidity and length of stay (unpubl. data).

We are acutely aware of the priorities, initiatives, funding, and incentive streams that may impact upon our partners, or our partnership, and where applicable to adapt as a system to the patient, population, and wider context. For example, while our partners occupy four of the top five positions nationally for low inpatient mortality (according to the 2011 NHS information center summary hospital mortality indicator scores) [10], all remain concerned to continually improve this outcome measure – as a reflection of their safety focus and the increasing driver of patient choice. We have set up and supported a learning program across the partnership, using best practice adapted to local context with the objective of identifying and intervening to support the deteriorating general ward inpatient, and so aim to further reduce inpatient cardiac arrests by 50% over 12 months. Some partners have already achieved this result.

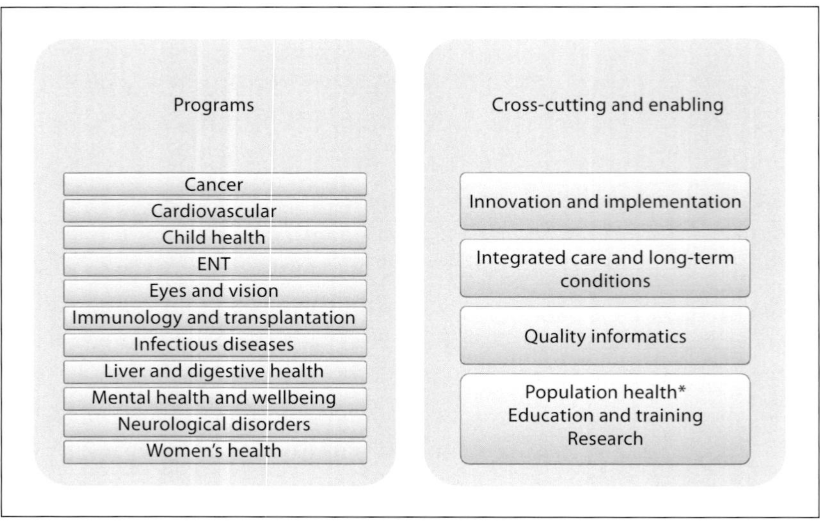

Fig. 2. Thematic programs and core support functions. *Population Health is both cross cutting support and a delivery program in its own right.

Collaboration

Collaboration is fundamental to our partnership model. It is embedded in our operating model, in the alliances we build, and in the transparent way in which we share information across our partners.

Thematic program directors (fig. 2) work with practitioners and experts within and beyond the system, and most crucially with patients and carers (care givers) to co-design and diffuse sustainable solutions. The program directors provide leadership through influence and support, with a 'warrant card' to do so where this enables delivery of specific approved projects across all partners. They do not directly control partner budgets or line-manage their staff, but work rather to facilitate and stimulate, empowered by the agreed partnership goals. Progress in programs towards agreed objectives informs decisions as to where resources should be directed. Scale is also critical to enable proper and timely evaluation of the impact of the type of healthcare interventions we wish to deliver. For example, through the leadership of the recently established Mental Health and Wellbeing program, the medical directors and a chief executive of our partner mental health trusts have agreed to cocreate and develop shared systems of measuring outcomes paired to resource use across key patient pathways and to share advances in service design aimed at improving value in mental healthcare. The scale of UCLPartners, bringing together already five major mental health providers (more than 11,000 clinical staff and 200 principal investigators) will be sufficient to power a proper study of the effect on value of outcomes reporting and performance feedback.

This process is aided by a central support team including experts in health improvement science, informatics, care quality and service integration, and local government and industry, working with the program directors to drive translational gap closure for individual projects targeted at population health and offer ready access to broad-ranging world-class academic expertise (fig. 2). As we engage across boundaries and build further care and research around patients' needs, traditional barriers are broken down. This may be promoted more easily within our partnership than through a traditional academic medical center as the projects and programs are based around specific goals rather than being responsible for administrative departmental systems. To enhance delivery, we are increasing capacity in implementation or 'improvement science' across the partnership and in conjunction with the other two London AHSCs. This is a rapidly evolving field that aims to bring an academically rigorous approach to the reliable delivery of care. The intellectual base of improvement science is broad, including disciplines as diverse as social sciences, economics, and engineering/process design (contained within the academic partners) – all of these have a bearing on the complex systems through which care is delivered to patients and adapted over time to maximize outcomes [11].

Building trusted alliances with patient and carer networks and organizations is fundamental to gain a deeper understanding of the problems within our population, and to co-design more relevant solutions. All programs are linked to appropriate local and national patient/carer organizations or charities. Having a patient voice embedded in our work enables the 'pulling' of innovations into practice at pace and scale.

It is similarly becoming clear that new forms of public-private partnerships will be required to develop the innovation space to allow creativity to thrive. These will need to accommodate, for example, new forms of commercial partnerships between pharmaceutical companies, the devices industry, and academia, enabling the sharing of expertise, risk and reward, and access to a much larger population (see Box 3). New academic-industrial partnerships are being forged, recognizing the need to engage more closely with concentrated expertise in fundamental and clinical science while drawing on industry's expertise in developing novel treatments, and acknowledging that the partnership provides an attractive scale of resources and population base for industry collaborations. International partnerships are being forged to bring together new approaches to common challenges – for example, as a partnership we have a formal collaboration with Yale and Yale New Haven Hospital in addition to the many bilateral academic linkages between research groups; and share a strong mutual interest with Partners in Boston to drive up value (outcomes that matter to patients per GBP or USD spent).

Box 3. London Cancer – an integrated cancer system

UCLPartners facilitated the co-design of 'London Cancer', an integrated cancer system serving our population. The co-design program involved a series of events, meetings and surveys to engage patients and patient organizations, staff across the care pathway, commissioners, and universities to agree on a shared vision and priorities with a skill-based rather than institutional board. This platform of engagement is now embedded within the system and has been used to define – for each cancer pathway – the measures that matter most to patients, with clear metrics for success (e.g. lives saved, % trial entry, enhanced patient experience). By demonstration of our working we recognize that 25% of all new cancer diagnosis occur through (late) presentation to the emergency room and carry a poor 1-year survival. We are treating each of these as a serious untoward incident undertaking root cause analysis along the whole patient pathway to progressively drive down this figure. The seamless integration of fundamental research, experimental cancer medicine, and a clinical trial research network presents an outstanding environment for cancer drug development. This results in year on year increases in recruitment to clinical trials and supports productive, collaborative partnerships with industry. We have also drawn on academic expertise from beyond biomedicine (e.g. humanities, implementation science, epidemiology) to better understand how to address the cultural challenge. A strong alliance has been established with the charity Macmillan Cancer Support, focused on strengthening meaningful engagement with patients, carers, and family doctors to improve patient experience and life with and beyond cancer ('survivorship') and to promote earlier diagnosis. Given these foundations, new alliances are being formed with large pharmaceutical companies to optimize opportunities for collaborative working at a system level.

Our collaborative efforts also include openly sharing performance information as a mechanism for driving the pace of implementation and diffusion of best practice across the partnership. Disseminating granular (i.e. at the level of the clinic, ward, surgery, pathway, or clinician) information, which compares performance among providers (locally, national, and internationally) can accelerate delivery through peer-to-peer support and reputational pressure, and also harnesses the moral purpose that motivates those who work in healthcare and the benefits of working within organizations where staff feel valued [12].

Exit Strategy

The fundamental aspects of the exit strategy are to embed and release new ways of working or innovations into clinical practice. Clearly, unless the partnership develops a means for flexible involvement during the lifecycle of a project, the core operating model of UCLPartners will be undermined by assumption of a growing management burden and the sustainability of change threatened. Furthermore, by creating new capacity across the system, we have the opportunity to stimulate further innovation across the partnership.

UCLPartners acts as an incubator for ideas to be cocreated, developed, and evaluated with focus, support, and under the scrutiny of the board and executive. Accordingly, a critical early step is enabling a partner(s) (within or outside UCLPartners) to adopt and adapt the transformation, with UCLPartners retaining a monitoring oversight and facilitating formal evaluation. This has already been achieved in stroke. Service consolidations across the partnership have been designed to create better quality and cost-effective models (e.g. the centralization of liver and pancreatic surgery, vascular surgery, neurosurgery, etc.), and specific informatics projects (e.g. new systems developed through the partnership with patients and industry to empower children and teenagers with asthma or diabetes using mobile phone and computer access to personalized health support, now supported by local partners/across populations).

For complex projects, such as new forms of integrated care, the transition from active support (providing advice when sought) to release will clearly run a longer time course, and will generate subsidiary projects running a separate time course and relevant to other contexts (see Box 4).

Box 4. Whittington Health

Whittington Health, an organization formed in April 2011 and covering acute and community provision for the population of Haringey and Islington in North London, is developing population-based models of care, focusing on patients with multiple comorbid conditions, with community services organized and delivered in partnership with primary care around the needs and preferences of individuals. Ongoing support and evaluation by UCLPartners will examine the ability of local models of care to achieve reduced reliance on hospital admission while improving the quality of care. We are working across the partnership on system-wide enablers. These include the development of a new tariff structure and enhanced information sharing. A particular challenge is to understand person-level provider costs across all settings of care to define new cohorts of the population, based not on individual diseases, but on the acuity of support needed from the system. Information sharing is pivotal to facilitate review of activity across the different settings of care. By linking systems, there is the opportunity to change how we monitor and manage performance, provide tools for risk stratification and proactive care, and improve communication across care teams, increasing patient safety and connectivity.

Conclusion

UCLPartners is a relatively new organization, but early experience suggests that it is a model that can deliver health and economic gain in a highly complex and fluid operating environment. Focus on delivery as well as new knowledge generation draws on and binds in a broader academic community, facilitating novel perspectives, and promoting a culture of innovation and evaluation.

Our experience to date suggests to us that a partnership model of the academic health science system is transferable. Self-assembly of the local alliance is important to promote adaptability and the inclusion of partners with a common ethos and aims. Many such local alliances are being formed across the country, which are now being developed formally through a nationally led system of academic health sciences networks that will provide the crucial framework for innovation and service transformation [13], supporting better value health care and wealth creation. These academic health science networks may also act as host sites for the interface with new educational structures and assist in better aligning multiprofessional education with the needs of patients, populations, trainees, and employers. Whereas the concentration of biomedical research in such networks will differ, all will be engaged in high-quality education, diffusion of innovation and evidence, and the conduct of applied health research – enhancing patient and population health gain and wealth creation nationally. Collectively, such an approach can help effect the cultural change that moves beyond reliance on central command and control that has characterized the NHS in previous years and fosters alignment of central priorities with the emergence of local ownership, clinical leadership, learning organizations, and an integrated system of care orientated around population health needs rather than professional groups or individual institutional priorities.

Disclosure Statement

UCLPartners has a commercial interest in the diabetic software interface referred to in the paper.

References

1 Dzau VJ, Ackerly DC, Sutton-Wallace P, Merson MH, Williams RS, Krishnan KR, et al: The role of academic health science systems in the transformation of medicine. Lancet 2010;375:949–953.

2 Green LW, Ottoson JM, Garcia C, Hiatt RA: Diffusion theory and knowledge dissemination, utilization, and integration in public health. Annu Rev Public Health 2009;30:151–174.

3 Daniels RJ, Carson LD: Academic medical centers – organizational integration and discipline through contractual and firm models. JAMA 2011;306:1912–1913.

4 Porter ME: What is value in health care? N Engl J Med 2010;363:2477–2481.

5 Grol R, Grimshaw J: From best evidence to best practice: effective implementation of change in patients' care. Lancet 2003;362:1225–1230.

6 Porter ME: Clusters and the new economics of competition. Harv Bus Rev 1998;76:77–90.

7 Stephenson J, Kuh D, Shawe J, Lawlor D, Sattar NA, Rich-Edwards J, et al: Why should we consider a life course approach to women's health care? 2011. http://www.rcog.org.uk/files/rcog-corp/14.10.11 SACLifecourse.pdf (accessed Jan 23, 2012).

8 Halligan A: The need for an NHS Staff College. J R Soc Med 2010;103:387–391.

9 Mountford J, Davie C: Toward an outcomes-based health care system: a view from the United Kingdom. JAMA 2010;304:2407–2408.

10 NHS Information Centre. Summary Hospital-level Mortality Indicator (SHMI) – Deaths Associated with Hospitalisation, England, April 2010 – March 2011, Experimental Statistics 2011. http://www.ic.nhs.uk/statistics-and-data-collections/hospital-care/summary-hospital–level-mortality-indicator-shmi/summary-hospital-level-mortality-indicator-shmi–deaths-associated-with-hospitalisation-england-april-2010–march-2011-experimental-statistics (cited Jan. 26, 2012).

11 Berwick DM: The science of improvement. JAMA 2008;299:1182–1184.

12 Finlayson B: Counting the smiles. London, UK, King's Fund 2002. http://www.kingsfund.org.uk/publications/counting_the.html (accessed Jan. 23, 2012).

13 Fish DR, Academic Health Sciences Networks in England. Lancet 2012, E-pub ahead of print.

Prof. David Fish, MD
UCLPartners Academic Health Science Partnership
170 Tottenham Court Road
London W1T 7HA (UK)
E-Mail david.fish@uclpartners.com

Alving B, Dai K, Chan SHH (eds): Translational Medicine – What, Why and How: An International Perspective.
Transl Res Biomed. Basel, Karger, 2013, vol 3, pp 18–28 (DOI: 10.1159/000343010)

The US Initiative: Clinical and Translational Science Awards – The UC Davis Perspective

Alice F. Tarantal[b–e] · Julie Rainwater[d] · Ted Wun[a,d,f] · Lars Berglund[a,d,f]

Departments of [a]Medicine, [b]Pediatrics, [c]Cell Biology and Human Anatomy, and [d]Clinical and Translational Science Center, [e]California National Primate Research Center, University of California Davis, and [f]VA Northern California Health Care System, Sacramento, Calif., USA

Abstract

The US Clinical and Translational Science Awards (CTSAs) have had a transformative impact on clinical and translational research and research education and training. As a member of the first cohort of funded CTSAs in 2006, the University of California Davis Clinical and Translational Science Center has taken full advantage of the mandate provided to achieve institutional transformation. To achieve these goals, a spectrum of initiatives was launched that challenged the existing institutional culture and resulted in the development of novel mechanisms and tools to engage trainees, investigators, institutional leaders, the community, and external partners. Examples of such initiatives include an expanded and integrated research education program, full leverage of institutional information technology resources, a versatile pilot project program, and the creation of an effective evaluation resource. Similar efforts throughout US CTSA institutions are contributing to expand resource utilization and efficiencies for national clinical and translational research capabilities.

The National Institutes of Health's (NIH) Clinical and Translational Science Award (CTSA) program has had a profound impact on the way clinical and translational research is being conducted throughout the nation [1–5]. This bold initiative, which was launched by the NIH in 2005, involved a thorough rethinking of the infrastructure support needed to conduct the innovative research required to serve future national needs. By providing a comprehensive framework with carefully considered guidelines, the NIH stimulated the nation's academic institutions to expand on previous familiar and well-tested mechanisms, and to move into a new frontier [1, 2]. Six years into this experience, it is apparent that much has changed at the institutional level, and in addition, national and regional transinstitutional collaborative efforts have expanded, supported by advances in reducing barriers to enhance the research agenda [5]. The University of California (UC) Davis was among the first 12 institutions to receive NIH funding for this award, and created the Clinical and Translational Science Center (CTSC) in 2006. This funding accelerated and further integrated an existing conscientious and careful planning effort for translational research to increase our institutional competencies, capabilities, and resources in this area [6, 7]. The development of the UC Davis CTSC has led to the development of new ways of bringing together a diverse faculty and facilitating research. Through these activities, the CTSC has had a substantial impact on virtually every area and infrastructure resource involved in promoting clinical and translational research at UC Davis, and has created a legacy that is likely to impact virtually every faculty and staff member and trainee [6–9]. Having been a part of the CTSA family since inception, we will focus on our experiences from the transformative process since this in many ways will likely mirror changes, adaptations, and gains experienced at many CTSA institutions [5, 6].

The Clinical and Translational Science Award as a Transformative Tool

The funding of a CTSA represented an important opportunity to rapidly accelerate clinical and translational research at UC Davis. At the time of funding, our institution had relatively modest resources in this area with a newly funded (2004) General Clinical Research Center, and a recently (2005) funded training program for clinical research (K30). Due to the timing of initiation of CTSA funding, the UC Davis CTSC was able to take full advantage of these recently established programs and rapidly integrate their activities into the CTSC. By capitalizing on the transformative mandate of the CTSC, we embarked on significant changes at virtually all levels, including the CTSC itself, the institution, those investigators conducting clinical and translational research, our expanding cadre of trainees, and ongoing community outreach efforts. The first years were characterized by an 'incubator' concept where seemingly disparate services and educational activities were brought together and integrated. By completing this concept, we were able to transform the CTSC into an efficient toolbox, used by both our faculty as well as others at our institution. Through this process, the CTSC developed into a coherent and creative team built on trust, mutual respect, and innovation, and this team spirit has been instrumental in creating a major cultural change by positively impacting our faculty, trainees, and institutional and community partners. Through an intense marketing effort, the CTSC has become the voice of change, opportunities, and initiatives. The majority of clinical and translational investigators at UC Davis have taken full advantage of the resources and opportunities provided through the CTSC, and have been carefully guided by CTSC program directors to the appropriate contacts, partners, and related mechanisms to enhance and facilitate their research [6]. Many faculty members have also independently approached the CTSC with ideas and suggestions, realizing that the CTSC provides a mechanism for change, innovation, and improvement.

Recognizing the need to achieve the goals outlined in the NIH CTSA initiative, the UC Davis CTSC has focused on integrating available resources to enhance and enrich scientific programs. Examples of such interactions across UC Davis include ensuring that trainees are part of the team for all funded pilot projects; using collaborative workshops and symposia to highlight pilot program thematic areas; involving the Biomedical Informatics program in creating a number of key, Web-based portals including a joint effort with the translational science leadership to develop a comprehensive and searchable CTSC facilities, cores, and resources website; initiating an online application for resource use; and implementation of the Pharmaceutical Assets Portal, which was part of a 2008 National Center for Research Resources (NCRR) supplemental award. Many of the CTSC programs have filled institutional voids and have grown to serve as important informational assets, such as the Evaluation and Biomedical Informatics programs. These and other programs also provide strong bridges to other CTSA institutions, making it possible to share and exchange best practices across CTSA sites to a much larger degree than was previously possible. Additional examples of the strong, integrative spirit of the CTSC is the collaboration between the Community Engagement and Education and Career Development programs to create core competencies for trainees in developing community outreach links. Conversely, our institution has enthusiastically embraced the opportunity provided by the CTSC to expand and grow critically important training and mentoring programs, and the CTSC educational strengths have successfully been leveraged many times in support of new training grant applications at UC Davis, which has stimulated cooperative and efficient resource sharing and the reduction of redundancies. This hallmark collaborative feature of the UC Davis CTSC is a common element that is integrated throughout all of the CTSC programs (key functions), and this has been instrumental in achieving and exceeding many of our original goals (fig. 1). Although all of the individual CTSC programs have contributed to the transformative impact, we focus below on four examples where institutional capabilities have been impacted in an unparalleled fashion.

Examples of Transformation

Pilot Projects

As much of the CTSA funding is provided in support of staff and trainees, the pilot project program represents an important flexible fund, critical to launch novel initiatives and to support high-risk research. In keeping with the spirit of the CTSC as a partnership program, and to provide an impetus for prospective pilot project applicants to use the full breadth of the CTSC program, we have closely linked our pilot program calls for proposals to ongoing translational activities. To achieve extensive

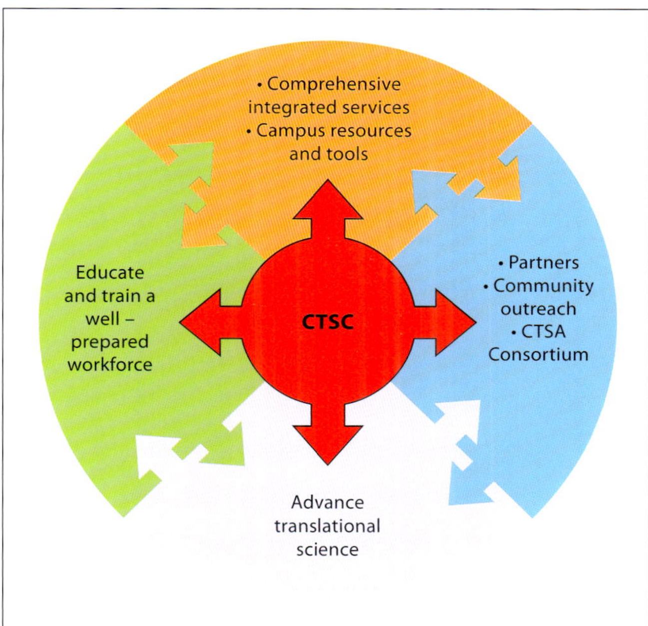

Fig. 1. The UC Davis CTSC has fully implemented the transformation of our biomedical research enterprise into an innovative, integrated, and collaborative 'open' academic home for clinical and translational research.

outreach, we have maximized and leveraged calls for pilot projects and planning grants by collaborating with many other centers and programs at UC Davis (table 1). Using this approach, the UC Davis CTSC pilot program has inoculated some key features, summarized below, with requirements that are targeted towards the formation of new multidisciplinary teams:

- Inclusion of trainees (e.g. undergraduate or graduate students, postdoctoral or clinical fellows)
- Active solicitation of partnerships with other UC Davis pilot project programs
- Linking CTSC pilot program announcements to translational workshops and symposia

CTSC pilot project funding has evolved into a structure focused on thematic areas and targeted to NIH initiatives in order to better serve investigators and contribute to the development of clinical and translational researchers of the future. All funded investigators and the title of their projects are highlighted on the CTSC website and quarterly progress reports and a final report are required from all awardees. The reporting structure requests that investigators include the overall outcome of the study in relation to the original goals; information on new biomarkers, tools, patents, or approaches transforming the area of study; trainees involved in the research; presentations and publications resulting from the supported study; and new grants or other funding opportunities (table 2). To date, the UC Davis CTSC has supported 100 pilot projects and planning grants spanning a wealth of

Table 1. Examples of UC Davis CTSC Pilot Project Partnerships

Partner	Subject of pilot call for proposals
School of Medicine	general call ('kick-off' of CTSC)
Alzheimer's Disease Center	neurodegenerative disorders
Children's Miracle Network	child health research; mentored pediatric resident research projects
UC Davis Comprehensive Cancer Center	cancer and inflammation
Center for Healthcare Policy and Research	linking social and medical sciences; comparative effectiveness research; electronic health records research
College of Engineering	senior undergraduate design capstone projects; early stage device development program
Medical Investigation of Neurodevelopmental Disorders (M.I.N.D.) Institute	neurodevelopmental disorders
California National Primate Research Center	translational investigational new drug (IND)-enabling studies
CTSA Consortium	translating regionalized healthcare delivery systems across CTSA sites
Center for Molecular and Genomic Imaging	in vivo imaging
Rosa B. Sherman Pediatric Research Fund	child health research

translational and clinical topics. Progress reports from these studies, as of May 2012, indicate 90 publications, the inclusion of 184 trainees, and funded grants totaling over USD 60 million (includes multiple types of NIH awards, Department of Defense grants, and a California Institute for Regenerative Medicine Disease Team grant) currently leveraging an approximate USD 1.1 million CTSC investment. To ensure cross-fertilization among investigators and trainees, the program has an annual retreat that shares widely research outcomes, and highlights new technologies, methodologies, therapies, and outreach to the community. A thematic focus is chosen annually (e.g. technology transfer, biorepositories) with a keynote speaker from another CTSA site. These retreats have been highly effective in team building and forming new partnerships, engaging faculty and their trainees in the CTSC program, and enhancing communications on crucial clinical and translational topics.

Table 2. Examples of UC Davis CTSC Pilot Project Outcomes

Project	Outcomes
Novel methods for defining infection and disease with *M. tuberculosis*	NIH new innovator award and NIH supplement 2 publications 1 trainee (medical student)
Detection of novel biomarkers of asthma and COPD in exhaled breath	3 private sector and federal awards 4 patents 16 publications 5 trainees (undergraduate and medical students, postdoctoral fellows)
Clinical evaluation of a fluorescence lifetime-based technique for demarcation of head and neck tumors: an intraoperative study in patients	NIH R21 (grant award) 2 publications 1 patent 8 trainees (graduate students, postdoctoral fellows, surgical residents)
Replacing the Schilling test: in vivo assessment of vitamin B_{12} absorption using ^{14}C-B_{12}	NIH Small Business Innovation Research (SBIR) award 7 publications 1 patent 1 trainee (graduate student)
The impact of human milk oligosaccharides and probiotics on the growth and intestinal microflora of premature infants	NIH R01 (grant) and NIH supplement award 8 publications 3 trainees (graduate students)
Novel hippocampal morphometrics for quantification of Alzheimer's disease structural correlates	NIH K01(training award) and private foundation award licensed new technology 5 publications 2 trainees (graduate students)
Human-computer interface for power wheelchair control	National Science Foundation (NSF) and foundation award 4 publications 1 patent 1 trainee (graduate student)
Animal models of vascular volume regulation	NIH R01 2 publications 1 trainee (graduate student)
The voltage-gated Kv1.3 channel in psoriatic disease: a novel therapeutic target	NIH R21 and private sector funding 1 publication patent licensed for psoriasis 2 trainees (MD/PhD student and veterinary surgery trainee)
Genetic modification of primary patient fibroblasts	W.M. Keck Foundation award 1 publication 1 trainee (postdoctoral fellow)

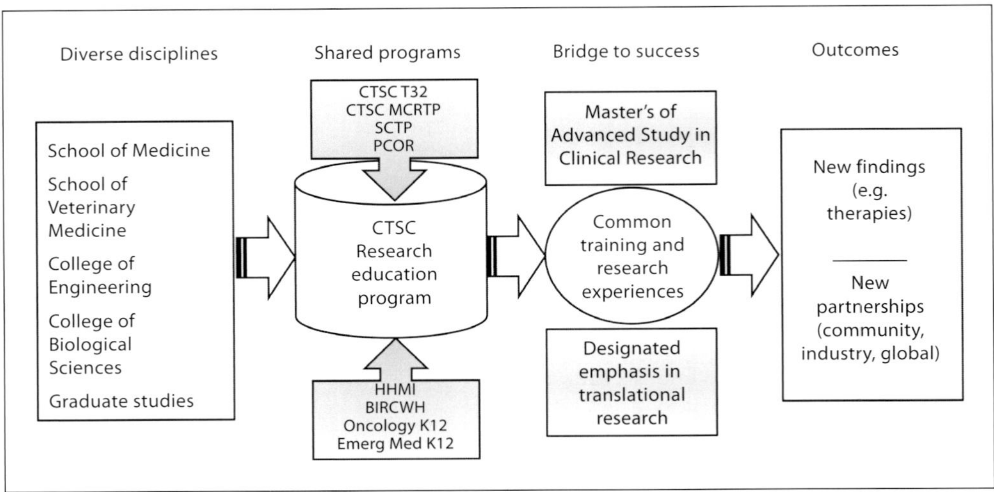

Fig. 2. The CTSC Education program attracts students from many schools and colleges and partners with a number of funded training grants. The program has two degree alternatives, a Master's of Advanced Study in Clinical Research and for graduate students, the option of a designated emphasis in translational research. MCRTP = Mentored Clinical Research Training Program; PCOR = Primary Care Outcomes Research program, funded by the Agency for Healthcare Research and Quality; SCTP = Stem Cell Training Program, funded by the California Institute for Regenerative Medicine; HHMI = Howard Hughes Medical Institute Integrating Medicine into Basic Science program, BIRCWH = NIH K12 program, Building Interdisciplinary Research Careers in Women's Health; Emerg Med = Emergency Medicine K12.

Research Education

The CTSC Education program has developed into a cornerstone of not only the CTSC but the institution as a whole. The overarching objective is to provide trainees a comprehensive, coordinated, and carefully planned experience with state-of-the-art multidisciplinary team training to ensure they become productive, critical thinking, highly trained, and well-rounded clinical and translational investigators [10]. Prior to the CTSC, training grants at UC Davis existed as isolated programs without any organized effort to share best practices. The CTSC took the initiative to integrate institutional training grants by leveraging common curricular features, creating a well-integrated community of scholars, and sharing tools for program evaluation [6]. Strategic integration is strongly evident from the coordination of a number of training programs beyond the CTSC, such as the Howard Hughes Medical Institute Integrating Medicine into Basic Sciences, California Institute for Regenerative Medicine Stem Cell Training Program, the NIH Building Independent Research Careers in Women's Health training program, the Primary Care Outcomes Research Program and two recently NIH-funded training (K12) programs in oncology and emergency medicine (fig. 2). These training grants target scholars at all levels from graduate students to junior faculty. This integrated portfolio of training grants now provides a well-utilized resource for a wide array of other institutional training programs with an interdisciplinary focus, such as those in the Colleges of Engineering and Biological Sciences [10, 11].

To break down institutional barriers, the CTSC Education Program established a training grants working group that meets quarterly to track program scholars, to address training program administrative issues, and develop recommendations for long-term solutions and process improvements. Since many of the education needs and programs at UC Davis overlap, integration of key elements of these programs bring together clinical and basic science faculty, graduate students, and medical students to work together as teams on classroom assignments, creating a supportive institutional culture for team science.

Biomedical Informatics

At UC Davis, the Biomedical Informatics program has developed into an invaluable and powerful resource that supports all other CTSC programs and a wide spectrum of investigators; this program has become an essential partner for all informatics activities at UC Davis. We have fully integrated the CTSC Biomedical Informatics program with institutional information technology resources in many areas, such as to provide support for the implementation of the Cohort Discovery software tool, powered by the Informatics for Integrating Biology and the Bedside (i2b2) data warehouse system, which was developed by the Partners system at Harvard University. Together with the University of Washington and UC San Francisco, UC Davis was awarded a CTSA research contract focused on informatics in 2008. Through this effort, researchers can now access large shared datasets of anonymized patient information across the three medical centers in order to design research studies and generate hypotheses [12]. Joining with the Partners program at Harvard University, where i2b2 was conceived, UC Davis continues to support these cross-institution research efforts. The project provides model policies and procedures for creating a shared, federated network of clinical research data among numerous institutions and harnessing the talents of many experts to focus on a particular problem. Through the CTSC, UC Davis has fully utilized the Cohort Discovery tool and is partnering with the four other UC CTSAs to achieve a UC-wide federated tool. Other areas of strength include national research data standards to reflect real-world requirements of the research computing environment [13]. In addition, by participating in nationwide CTSA consortium efforts, the program has brought many other valuable tools to UC Davis, including REDCap (a volunteer registry and investigator profile resource), which are widely used by our investigators [14, 15].

Evaluation

Prior to CTSC funding, research and training grant evaluation activities were pursued sporadically. Recognizing the need for measureable outcomes of success, a more rigorous approach to evaluation was undertaken as the CTSC was implemented, and resulting from the CTSC accelerator 'effect', this effort has expanded several-fold. Combined with institutional resources, the CTSC has transformed the Evaluation program from a loose association of research personnel with disparate skills and interests into a well-functioning integrated unit that rigorously evaluates the diverse CTSC activities as well as those of other large, complex research programs throughout UC Davis. In doing so, Evaluation program staff have helped to articulate goals and develop tools that continually and accurately measure progress and outcomes. Working closely with the Biomedical Informatics team, many tracking functions have become automated through the use of the online application for resource use or productivity tracking tools developed for specific program areas. For example, data regarding the research progress of CTSC trainees and pilot project awardees are collected and stored in a custom database that can interface directly with software used to visualize the multidisciplinary nature of their research using social network analysis. As an integral part of the effort to continually improve the CTSC, the Evaluation program has undertaken focused studies that describe utilization and satisfaction with CTSC resources and pilot project funding. A key success has involved expanding the evaluation efforts relevant for the various CTSC education programs which allows for real-time feedback from faculty and trainees on the curriculum and a schedule of Web-based surveys to track the careers of program alumni. The Evaluation team has also been critical in securing support for cross-program studies, including an NIH administrative supplement with the CTSC Community Engagement program to inventory community-based research partnerships and conduct a series of interviews and surveys to determine potential collaborations that benefit both the CTSC and the community.

The University of California Biomedical Research Acceleration, Integration, and Development Initiative

As a direct result of CTSA funding, the five UC CTSAs (Davis, Irvine, Los Angeles, San Diego, and San Francisco) have initiated a broad-based collaboration to expand institutional integration, using the respective CTSA institutional programs as the organizing principle. Several workgroups have been launched aiming to reduce barriers and share advances in institutional review board review processes, research contract negotiations, sharing deidentified electronic health records, and developing capabilities for drug and device development. Progress, to date, includes a memorandum of understanding to allow reciprocal use of institutional review boards across the UC system, many new master contract agreements, and federated use of the Cohort Discovery tool. These advances reflect a strong regional impact of the CTSA programs and provide a robust example of cross-institutional efforts to achieve efficiencies and reduce redundancies to fully capitalize on an economy of scale throughout the translational spectrum.

Conclusion

The CTSC has had a transformative impact at UC Davis, and has markedly improved the ease and ability of UC Davis investigators to engage in clinical and translational research. Some of the most transformative changes include the creation of a recognized, highly visible, academic home for clinical and translational research; the catalyst of novel inter- and multidisciplinary research projects; support for development and application of novel tools and technologies; efforts to streamline, consolidate, and amplify services supporting clinical and translational research; a strong regional impact; and the creation of an experienced, skilled, and well-trained biomedical workforce ready to address translational research challenges. Similar achievements are seen at virtually all CTSA institutions; the integrated effect of this change provides a solid foundation for continuing efforts to improve human health and advance biomedical research efficiently across the nation [16–18].

Acknowledgement

The authors gratefully acknowledge the support from the National Center for Advancing Translational Sciences (TR000002).

References

1 Zerhouni E: Translational and clinical science – time for a new vision. N Engl J Med 2005;353:1621–1623.
2 Zerhouni EA, Alving B: Clinical and Translational Science Awards: a framework for a national research agenda. Transl Res 2006;148:4–5.
3 Califf RM, Berglund L, for the CTSA Principal Investigators: Linking scientific discovery and better health for the nation: the first three years of the NIH's Clinical and Translational Science Awards. Acad Med 2010;85:457–462.
4 Reis SE, Berglund L, Bernard GR, Califf RM, FitzGerald GA, Johnson PC: Reengineering the National Clinical and Translational Research Enterprise: the strategic plan of the National Clinical and Translational Science Awards Consortium. Acad Med 2010;85:463–469.
5 CTSA Principal Investigators: Preparedness of the CTSA's structural and scientific assets to support the mission of the National Center for Advancing Translational Sciences. Clin Transl Sci 2012;5:121–129.
6 Asmuth D, Wun T, Mullen N, Garcia E, Chedin E, Whitney E, Gillis M, Anderson K, Tarantal A, Kenyon N, Curry FR, Berglund L: UC Davis CTSA: coming of age. Clin Transl Sci 2009;2:98–101.
7 Berglund L, Tarantal A: Strategies for innovation and interdisciplinary translational research: removal of barriers through the CTSA mechanism. J Investig Med 2009;57:474–476.
8 Kasim-Karakas S, Hyson D, Halsted C, van Loan M, Chedin E, Berglund L: Translational nutrition research at UC Davis – the key role of the Clinical and Translational Science Center. Ann NY Acad Sci 2010;1190:179–183.
9 Lee JS, Bertakis K, Meyers FJ, Chedin E, Tarantal A, Anderson K, Berglund L: Cardiovascular disease in women – challenges deserving a comprehensive translational approach. J Cardiovasc Transl Res 2009;2:251–255.
10 Meyers FJ, Begg MD, Fleming M, Merchant C: Strengthening the career development of clinical translational scientist trainees: a consensus statement of the Clinical and Translational Science Award (CTSA) Research Education and Career Development Committees. Clin Transl Sci 2012;5:132–137.
11 Domino SE, Bodurtha J, Nagel JD, the BIRCWH Program Leadership: Interdisciplinary research career development: building interdisciplinary research careers in Women's Health Program best practices. J Women's Health 2011;11:1587–1601.

12 Anderson N, Abend A, Mandel A, Geraghty E, Gabriel D, Wynden R, Kamerick M, Anderson K, Rainwater J, Tarczy-Hornoch P: Implementation of a deidentified federated data network for population-based cohort discovery. J Am Med Inform Assoc 2012;19:e60–e67.

13 Tenenbaum JD, Whetzel PL, Anderson K, Borromeo CD, Dinov ID, Gabriel D, Kirschner B, Mirel B, Morris T, Noy N, Nyulas C, Rubenson D, Saxman PR, Singh H, Whelan N, Wright Z, Athey BD, Becich MJ, Ginsburg GS, Musen MA, Smith KA, Tarantal AF, Rubin DL, Lyster P: The Biomedical Resource Ontology (BRO) to Enable Resource Discovery in Clinical and Translational Research. J Biomed Inform 2011; 44:137–145.

14 Harris PA, Taylor R, Thielke R, Payne J, Gonzalez N, Conde JG: Research electronic data capture (REDCap) – a metadata-driven methodology and workflow process for providing translational research informatics support. J Biomed Inform 2009;42:377–381.

15 Harris PA, Scott KW, Lebo L, Hassan N, Lightner C, Pulley J: ResearchMatch: a national registry to recruit volunteers for clinical research. Acad Med 2012;87: 66–73.

16 Collins FS: Reengineering translational science: the time is right. Sci Transl Med 2011;3:90cm17.

17 Skarke C, FitzGerald GA: Training translators for smart drug discovery. Sci Transl Med 2010;2:26cm12.

18 Dilts DM, Rosenblum D, Trochim WM: A virtual national laboratory for reengineering clinical translational science. Sci Transl Med 2012;4:118cm2.

Lars Berglund, MD, PhD
Department of Medicine, University of California, Davis UC Davis Medical Center, CTSC
2921 Stockton Blvd, Suite 1400
Sacramento, CA 95817 (USA)
E-Mail lars.berglund@ucdmc.ucdavis.edu

Alving B, Dai K, Chan SHH (eds): Translational Medicine – What, Why and How: An International Perspective.
Transl Res Biomed. Basel, Karger, 2013, vol 3, pp 29–35 (DOI: 10.1159/000343024)

National Agenda and Strategies for the Development and Implementation of Translational Medicine

Yupei Zhao

Peking Union Medical College Hospital (PUMCH), Beijing, PR China

Abstract

As the country with the world's largest population, China is facing great challenges in managing chronic noncommunicable diseases, prevention, and control of major infectious diseases and coping with the health and medical care of an aging population. Hence, there is an urgent need for the development of new approaches based on scientific discoveries emerging from medical research. The application of translational research has enormous potential for addressing these challenging clinical needs. Fundamental research applied to drug discovery and delivery, biomarkers for medical decision-making, and device development are just three areas which are showing significant growth potential in advancing the care of patients on a large scale. China has unique advantages in translational medical research. For instance, there are plenty of patients with diseases in need of new approaches to prevention and treatment. Moreover, because the Ministry of Health (MOH) of the People's Republic of China has a major role in mandating treatment practices in more than 90% of hospitals in China, there is an integrated managerial approach for building an efficient platform of translational medical research. China will take the strategies of medical discovery implementation, fundamental platform construction, national resource integration, and research group management to progressively establish a national biobank of clinical resources and the national centers of translational medicine. The goal is to thoroughly integrate national resources, and achieve complementary collaboration to form an 'application-education-research synchronized system' for an independent, innovative, and sustainable model linked to the development of translational medical research.

Current State of Health and Medical Care Industry in China: Demand and Trends

China makes up 1/5 of the world's population. Consequently, its health and medical care play a crucial role in the development of global medical care. While China has achieved great progress in the past three decades, it still faces tremendous pressure and many challenges in health and medical care.

First, China is confronted with an urgent need for disease prevention in cardiovascular and cerebrovascular diseases, cancer, diabetes mellitus, and other chronic noncommunicable diseases – all major causes of morbidity and mortality for Chinese residents. The increasing prevalence of these diseases constitutes a heavy medical burden. In addition, the incidence of potentially fatal infectious diseases, such as tuberculosis, AIDS, and viral hepatitis, remain at a high level. Moreover, new infectious diseases, such as SARS and influenza A (H1N1), are emerging constantly. China still faces the daunting prospect of optimal prevention and control of these common diseases [1].

Concomitant with the challenge of managing population growth, the related issue of an aging population is emerging with increasing importance. Disease prevention and implementation of health services have brought great economic pressures on individual households and society in general. Therefore, in order to meet the increasing and urgent needs for health and medical care, China must on a national level break down the barriers among clinical care, fundamental research, and drug discovery by using the research framework of translational medicine within the context of its special national needs and resources. Moreover, China must identify mechanisms for integrating its own unique 'application-education-research' system, focusing on bridging clinical demands with drug discovery that will benefit the health and improve the medical care of Chinese people.

Background of Translational Medical Research in China: Bottlenecks and Advantages

The concept of translational medicine was introduced to China in 2005. During the past 7 years of development, considerable funds have been invested in patient-oriented medical research projects, resulting in a significant increase in research discoveries that have gained national recognition among peers; however, meaningful practical achievements are rare. Thus, an overall conclusion could be drawn that the process of translational medical discovery and implementation needs substantial improvement in China, and that a significant gap between fundamental research and clinical application persists.

In finding solutions to these perplexing bottlenecks, one must first account for the modest success of previous attempts at developing translational medical research. At the heart of the matter, China has not yet demonstrated the capacity for independent innovation; in other words, most fundamental medical research has been either confirmatory or an incremental extension of European and American discoveries. Therefore, China needs to demonstrate the ability to perform innovative fundamental research and to translate these discoveries into health care more rapidly. Secondly, resources are not well integrated, particularly since there are a number of translational medical research institutes in China. Mainly focusing on a specific disease, these institutes are handicapped by their limited scope and depth; as a result, they are unable to organize and manage interdisciplinary collaboration and large-scale research. To some extent, this situation has led to wasted resources and inordinately slow research progress. Furthermore, the maldistribution of personnel has resulted in the suboptimal function of the entire research organization. Sharing and appropriately distributing basic resources for translational medical research need to be improved substantially.

China, however, has its own unique advantages in national resources:

1 The large base number of the Chinese population provides sufficient numbers of patients for comprehensive clinical trials that can be conducted more rapidly and, thus, result in discoveries that would require a much longer time frame than in countries with substantially smaller populations. Moreover, particularly complex cases are evaluated and treated in a small number of national centers, thus providing a critical mass for translational research; the results of studies from these centers should lead to improved patient management. Progress in advancing treatment for medical diseases and conditions is usually slowed considerably by the inability to study these problems in an adequate number of patients, resulting in a scarcity of tissue and blood samples for analysis, and underpowered studies, such as those that address the efficacy of new treatments. These national centers provide a major advantage by promoting a critical mass of patients for efficient and rapid translational research. Additionally, because the Chinese race is typical among the Asian ethnic groups, studies involving Chinese patients have great potential value for advancing global translational medical research.
2 One of the major goals of the Chinese government in recent years is to ease the pressure of medical demands, optimize medical services, and improve health conditions. In the past decade, great attention and commitment by the government has led to a remarkable rise in investment in medical research and support for the development of translational medical research in China.
3 The management system for the Chinese medical industry is conducive to concentrating and integrating national medical research, thus providing complementary advantages for conducting systematic and comprehensive research.

National Strategies for the Development of Translational Medicine

China has published a strategy for the construction of an innovative mechanism for developing national medical technology. Thus, vertical and horizontal integration among the life sciences and medical and pharmaceutical sciences with major disease areas such as cancer, cardiovascular and cerebrovascular diseases, diabetes, viral hepatitis, and other severe diseases are the highest priority for translational research [2, 3].

For the above reasons, the recent translational medical research strategy is divided into four levels:

1 Implementation of research discoveries: based on abundant patient resources as well as clinical teams trained for testing advances from the laboratory, translational research centers will actively seek collaborations internationally with investigators and organizations that are leading appropriate, mature translational research projects. The particular foci will be on prevention and treatment of cardiovascular and cerebrovascular diseases, malignant tumors, diabetes, severe infectious diseases, and other diseases with substantial morbidity and mortality.
2 The construction of a basic platform: building a national standard for management and integration of clinical resource samples, electronic medical records, and health research data in order to support translational medical research. Subsequent steps for moving forward will be the formation of national mechanisms for sharing information and other resources [4–9].
3 The integration of national resources: since 2009, more than 40 translational medical centers have been established. However, overlapping research goals and low efficiency have resulted in a waste of resources and hampered progress. It is necessary for China to implement a national plan with consensus approaches to unite advanced institutions in building national translational research centers [10].
4 Training: China will invite international research leaders to conduct relevant translational medical research in China. This will include exchange of projects and programs at a national level and will involve training of interdisciplinary research personnel, such as research doctors and nurses, basic research assistants, infectious disease epidemiologists and biostatisticians, research management coordinators, etc. Interdisciplinary teams or translational medical teams can then be developed [11, 12].

China will implement a 10-year plan to enhance the transformative power of scientific research to form an 'application-education-research system' in an independent, innovative and sustainable model for the development of translational medical research. Along with the development of translational medicine, the state investment in translational medicine research will grow rapidly. Success will gradually broaden funding sources, with joint investment from social enterprises, institutions, and charities to form a mature model of translational medical research [13, 14].

Blueprint for National Centers of Translational Medicine

Currently, China faces a precarious healthcare future because of its vast geography and the size of its population, which is the largest of any country in the world. Therefore, to meet these exceptional challenges not faced by Western countries, creative new solutions are required. To meet this goal, China will establish two national translational medicine development centers: one in the North and one in the South.

Beijing, the site for the northern national translational medical research center, will integrate the superb scientific resources of the Chinese Academy of Sciences and Chinese Academy of Medical Sciences. The faculty of Peking Union Medical College Hospital, the first-ranking hospital in China which is linked to the Chinese Academy of Medical Sciences, will be the core institution for the national northern center. Based on the abundant resources in talent and technology, the northern translational research will encompass numerous disease domains, including cancer, diabetes, chronic noncommunicable diseases, serious infectious diseases such as AIDS, tuberculosis, viral hepatitis, immune system disorders, and autoimmune diseases. The northern translational medical research center will focus on pathogenesis and provide an effective model for disease prevention and treatment, particularly through new drug and vaccine development.

The southern center, located in Shanghai, is based at Shanghai Jiao Tong University because of the academic advantages of the large number of institutes affiliated with it. The Shanghai government has granted this new institution, which was founded in 2010, a large amount of financial and space resources with the expectation that the investments will improve the treatment of major diseases, such as malignancy, metabolic disorders, and cardiovascular and cerebrovascular diseases. The Shanghai model of translational medical research in China will be studied by other qualified medical universities in the southern provinces.

A smoothly operating mechanism that integrates fundamental research with patient health care will efficiently expedite implementation of research into clinical applications. It will establish a priority for disease prevention, moving it forward to the early stages of diseases (e.g. biomarker discovery for early-stage diagnosis and effective individualized treatment). With the valuable experiences gained from the translational system, the research goals of China will gain in scope and depth.

The strategy of translational medical research in China will encourage development of fundamental medical research and interdisciplinary research equally. Thus, these plans are expected to substantially increase achievements and establish a model for clinical and basic research training that is focused on a shared platform of research and technology, and on strengthening integration and innovative team building among diverse hospitals. In conclusion, China aims to establish a highly integrated, comprehensive, and Web-connected translational medical research system [15, 16].

Several of the government's goals for promoting the development of translational medicine and research will also facilitate cooperation with the new drug development industry in China. The government plans to build a comprehensive biobank, mobilizing and integrating all local government and community resources. The national biobank is made up of human specimens such as blood, urine, cerebrospinal fluid, and fresh and paraffin-embedded tissues; these specimens are obtained for research purposes from patients undergoing diagnostic, therapeutic, and surgical procedures [17, 18]. Initially, the specimens will be obtained from patients with four major types of diseases: malignancy, cardiovascular, and neuropsychiatric and metabolic diseases, as well as from patients with rare diseases, where it is difficult for a single investigator to collect a sufficient number of specimens to make meaningful observations. Subsequently, tissue collection will be extended to healthy people in different age groups. The bank will provide standardization of samples for new drug development and research on major disease mechanisms. The hospitals plan to make full use of specimen resources to build tissue bank-centered comprehensive public service platforms and work with advanced pharmaceutical companies and research institutions to develop new drugs. The clinical biobank program will link the translation of basic scientific research and clinic application effectively [19–22].

To summarize, the MOH and the Chinese government will lead the way in national planning, resource sharing, integrating biospecimen storage, developing multiple collaborative research efforts, and fostering an 'application-education-research system' program, thus shaping an innovative and sustainable model for the development of translational medical research in order to fulfill the specific and unique needs of the people of China.

References

1 The '12th Five-Year Plan' of Medical Science and Technology Development, No. 552. Beijing, Ministry of Science and Technology.

2 National Long-Term Science and Technology Development Outline (2006–2020). Beijing, PRC State Council, 2006.

3 Chinese Medical Science and Technology Development Report 2012, Chinese Academy of Medical Sciences. Science Press.

4 Murphy SN, Dubey A, Embi PJ, et al: Current state of information technologies for the clinical research enterprise across academic medical centers. Clin Transl Sci 2012;5:281–284.

5 Pulley JM, Harris PA, Yarbrough T, et al: An informatics-based tool to assist researchers in initiating research at an academic medical center: Vanderbilt customized action plan. Acad Med 2010; 85:164–168.

6 Ectors N: International and national initiatives in biobanking. Verh K Acad Geneeskd Belg 2011;73: 5–40

7 De Paoli P: Institutional shared resources and translational cancer research. J Transl Med 2009;7:54.

8 Bell WC, Sexton KC, Grizzle WE, et al: Organizational issues in providing high-quality human tissues and clinical information for the support of biomedical research. Methods Mol Biol 2010;576:1–30.

9 Herpel E, Koleganova N, Schreiber B, et al: Structural requirements of research tissue banks derived from standardized project surveillance. Virchows Arch 2012;461:79–86.

10 Dai K: The development strategy of translational research based on China's current national situation; in Sanders S (ed): Selected Presentations from the 2011 Sino-American Symposium on Clinical and Translational Medicine. Washington, Science AAAS, 2011, p 7.

11 Hobin JA, Galbraith RA: Engaging basic scientists in translational research. FASEB J 2012;26:2227–2230.

12 Westfall JM, Ingram B, Navarro D, et al: Engaging communities in education and research: PBRNs, AHEC, and CTSA. Clin Transl Sci 2012;5:250–258.

13 Reis SE, Berglund L, Bernard GR, et al: Reengineering the national clinical and translational research enterprise: the strategic plan of the National Clinical and Translational Science Awards Consortium. Acad Med 2010;85:463–469.

14 Government, academy and venture firms come together to fund translational and early-stage development. Translation Med J 2012;1:104–106.

15 Marantz PR, Strelnick AH, Currie B, et al: Developing a multidisciplinary model of comparative effectiveness research within a clinical and translational science award. Acad Med 2011;86:712–717.

16 Linking practice-based research networks and Clinical and Translational Science Awards: new opportunities for community engagement by academic health centers. Chinese Southern Translational Medicine Forum, 2011.

17 National Cancer Institute Best Practices for Biospecimen Resources (National Cancer Institute National Institutes of Health U.S. Department of Health and Human Services). Rockville, National Cancer Institute Office of Biorepositories and Biospecimen research, 2012.
18 Biobank standard of national association of pharmaceutical biotechnology (trial). Chinese Pharmaceutical Biotechnology 2011;1:71–79.
19 Payne PR, Huang K, Keen-Circle K, et al: Multidimensional discovery of biomarker and phenotype complexes. BMC Bioinformatics 2010;11(suppl 9):S3.
20 Zerhouni EA: US biomedical research: basic, translational, and clinical sciences. JAMA 2005;294:1352–1358.
21 Vassal G, Borella L, Pierre A, et al: Translational research and cancer plan. Bull Cancer 2007;94:1107–1111.
22 Hawk E, Viner JL: What is the future of oncology? National Cancer Institute initiatives to improve research, development, and implementation in cancer prevention and treatment? Semin Oncol 2006;33(6 suppl 11):S6–S9.

Yupei Zhao, PhD
Peking Union Medical College Hospital (PUMCH), Translational Medicine Center of PUMCH
#1 Shuaifuyuan Road, Dongcheng District
Beijing, 100730 (PR China)
E-Mail zhao8028@263.net

Alving B, Dai K, Chan SHH (eds): Translational Medicine – What, Why and How: An International Perspective.
Transl Res Biomed. Basel, Karger, 2013, vol 3, pp 36–46 (DOI: 10.1159/000343028)

Developing a Translational Research Workforce in the USA

Thomas A. Pearson[a] · Wishwa N. Kapoor[b] · Robert G. Holloway[a]

[a]Clinical and Translational Science Institute, University of Rochester Medical Center, Rochester, N.Y., and
[b]Institute for Clinical Research Education, University of Pittsburgh Medical Center, Pittsburgh, Pa., USA

Abstract

The training of a workforce with translational research skills has been an essential part of the Clinical and Translational Science Award Program of the NIH since 2006. This training has emphasized broad multidisciplinary skills to address translational research problems but also a focus on one part of the translational research spectrum. Didactic training is necessary and should be competency-based but should evolve with the rapidly developing methods of this field. Multidisciplinary mentors are especially important and the development of high quality research mentors is a good long-term investment. Clinical and translational research education and training programs frequently involve several scientific disciplines and therefore should be coordinated and resourced at the institutional level. A solid commitment to training with these characteristics will be needed to fill the current void of translational scientists in the U.S. and other research-intensive countries.

A crisis in clinical research in the USA which had been building for decades came to a climax in the mid-1990s, as the proportion of National Institutes of Health (NIH) extramural research funds allotted to clinical and translational research shrank, resulting in fewer physician-scientists applying for grants in patient-oriented research. Only 10% of NIH funds around 1994 supported 'hands on' clinical research with direct patient involvement [1–6]. In 2005, NIH Director Elias Zerhouni announced the launching of the Clinical and Translational Science Award Program (CTSA) [7], designed in part to address the shortage of clinical and laboratory-based biomedical researchers involved with translation of basic discoveries to human applications, i.e. development of drugs, devices, and diagnostic tests, their implementation into clinical use, and the dissemination of these products to the clinical practice and community levels. This emphasis on training and career development has continued under the direction of NIH Director Francis Collins, who sees 'reinvigorating and empowering

Table 1. Seven premises supporting the rationale for development of a US translational research workforce

1.	A broad spectrum of training provides the multidisciplinary skills needed to address contemporary clinical/translational research problems
2.	Clinical and translational research training requires focus on one part of the clinical/translational research spectrum
3.	Didactic training needs continued evolution of coursework in the theory and methods of translational research
4.	Clinical/translational research training should be competency-based to impart practical research skills
5.	Effective clinical and translational research training involves a multidisciplinary team of co-mentors
6.	Development of high-quality research mentors is crucial to expanding translational research
7.	Clinical and translational research educational programs should be coordinated and promoted at the institutional level

the biomedical research community, including new scientists, as one of five strategic priorities for re-engineering of translational science at the NIH [8].

In discussing the US experiences in developing new programs to quantitatively and qualitatively improve the translational research workforce, we list seven premises which have supported these efforts and have helped to organize didactic, mentored research, and career development programs at the institutional and national levels (table 1).

Seven Basic Premises Underpinning the Development of the Translational Research Workforce in the USA

A Broad Spectrum of Training Provides the Multidisciplinary Skills Needed to Address Contemporary Clinical/Translational Research Problems

Most careers in the life sciences begin with the building blocks of biology, physics, chemistry, and mathematics, and then become increasingly focused on the theory and methods of the chosen scientific discipline. Translational science, which seeks to bridge basic research to clinical application, requires knowledge and skills across an even broader spectrum (fig. 1). For example, several of the steps required for the 're-engineering of translational science' [8], such as therapeutic target validation, virtual drug design, and biomarkers, will depend on investigators with a firm understanding of these methods and a good grasp of their application. The principles and practice of experimental therapeutics can include device and prototype development, preclinical toxicology, efficacy testing in human cells or other models, phase zero (first-in-human) clinical trials, and phase one (early learning) clinical trials. These require not only knowledge of human pathophysiology and pharmacology, but also

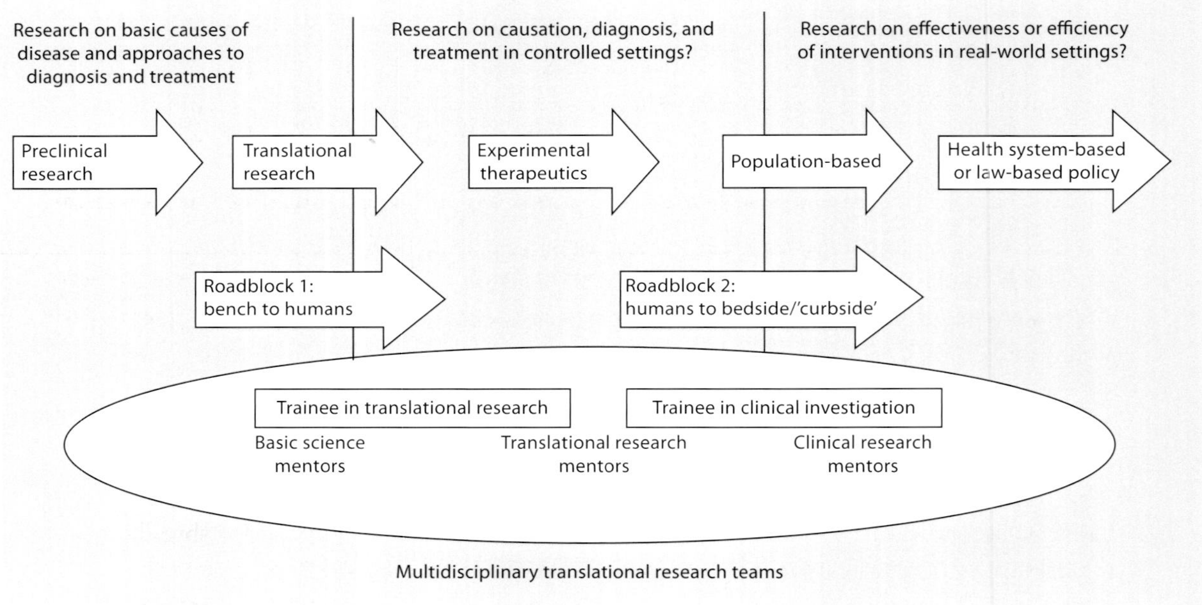

Fig. 1. Conceptual rationale for the Rochester Clinical and Translational Science Institute education and research key function.

aspects of clinical medicine and human subjects' protection. The development and efficacy testing of drugs, devices, and diagnostics with clinical trials depends not only on regulatory science required for approval, but also on clinical trial design, which increasingly includes health economics, patient satisfaction/quality of life, and other psychosocial measures. Finally, the dissemination of the application to practices and communities will likely involve comparative effectiveness research, implementation science, and postmarketing epidemiologic surveillance.

The recognition that the pipeline of discoveries was impeded by the lack of translational scientists [5, 7] then brings the question: 'What is the best strategy to fill this workforce gap between basic scientists and clinicians?' Clearly, research workforce analyses have identified an oversupply of basic scientists trained in North America, Europe, and Asia, without sufficient positions to employ them all. Concurrently, clinicians (MDs, DDS, etc.) have witnessed increased pressure on their time and effort due to a shortage of care providers, reduction in clinical revenues, and increased administrative requirements. This has led to reduced opportunities for exposure to research training in medical school and graduate medical education. In this setting, three options are obvious. First, basic scientists could receive additional training in clinical and translational science to apply their basic knowledge and methods to translational research. Second, clinicians might receive translational and clinical science training that encourages them to apply their clinical acumen to translational research. Third, training programs may develop a de novo type of investigator with the basic, translational, and clinical skills required. With funding from the NIH and foundations, a number of PhDs in translational science have been trained, although it is too early to evaluate their effectiveness in alleviating the shortage of translational scientists.

Clinical and Translational Research Training Requires Focus on One Part of the Clinical/Translational Research Spectrum

One potential risk in the education of translational researchers is to graduate trainees who have some knowledge of a broad spectrum of disciplines but are not experts in any one discipline. The ideal graduate might be considered the 'T-shaped' scientist [9] with some breadth of understanding of translation but also depth of expertise and knowledge related to a more narrow type of translational research, in order to contribute novel and highly technical studies to the research community and its sponsors. The clinical and translational science programs at both the University of Pittsburgh and the University of Rochester addressed this problem by developing four translational 'tracks' in which to focus personnel and resources in specific areas of translational science. The Pittsburgh CTSA program focuses on four methodological areas: T1 translation, clinical trials, health sciences research, and comparative effectiveness research. Likewise, the Rochester CTSA has four centers, each with a methodological focus: experimental therapeutics, patient-oriented research (as defined by the NIH), comparative effectiveness research, and community-based participatory research [10]. As shown in table 2, each of these centers varies with the expertise they provide, with some overlap, integration, and coordination among centers. They also require personnel with differing skill sets, as well as specific types of facilities and equipment. Linked to each of these methodological foci are educational programs, usually master's degrees, requiring one year of didactic coursework and one or more years of mentored research experience. Each program shares a variety of research skill workshops, as well as a core of methodological coursework in areas or fields such as epidemiology, biostatistics, and the responsible conduct of research. However, each program also has its unique emphasis. For example, Rochester's Master's in Translational Science emphasizes courses on human pathophysiology, experimental therapeutics, and translational laboratory technologies. The Master's of Science in Clinical Investigation emphasizes advanced epidemiologic and biostatistical methods, the design and conduct of clinical trials, and social and behavioral medicine. The programs in comparative effectiveness emphasize health services research, including decision analysis, cost-effectiveness, and implementation science. The Master's in Public Health includes health policy, community engagement, environmental health, and preventive medicine.

Each of these programs requires a scholarly project of quality consistent with publication in a peer-reviewed journal. The trainee's master's thesis then provides the opportunity to focus on a specific topic, providing the trainee with a mentored research

Table 2. The spectrum of translational science addressed by centers of excellence in the Rochester Clinical and Translational Science Institute [10]

Center programs and functions	Center for Human Experimental Therapeutics (CHET)	Clinical Research Center (CRC)	Center for Research Implementation and Translation (CRIT)	Center for Community Health (CCH)
Translational science	experimental therapeutics/drug development	patient-oriented research	comparative effectiveness/ implementation research	community-based participatory research
Expertise/key roles	interface with basic scientists; phase I trials; first in human studies; regulatory support; clinical trials coordination; clinical pharmacology	phase II, III trials; metabolic studies; imaging, exercise science; protocol development and review; high-risk studies	phase IV trials; cost effectiveness; outcomes research; comparative effectiveness; implementation science; mixed and novel methods	population-based research; community-based studies; cultural competency; community engagement; community navigator; practice-based research
Personnel/ staff	clinical pharmacologists; clinical trial managers; biostatisticians; bioinformatics; regulatory scientists	clinical research nursing; nutritionists; exercise physiologists; behavioral scientists; research coordinators	cost-effectiveness experts; health economists; clinical decision analysts; ethnographers; complex systems researchers; implementation scientists	health educators; epidemiologists; behavioral scientists; preventive medicine/ public health physicians
Facilities/ equipment	research management software; biomedical informatics; clinical trial coordination center; clinical materials support unit; GMP facility; basic science core laboratories	inpatient and outpatient research clinic; exercise laboratory; imaging and anthropometric equipment; behavioral assessment lab; clinical laboratory	administrative databases; electronic medical records; data warehouses; practice-based research network clinics	practice-based research network clinic; community-based clinical research center; community registries; databases; vital statistics; regional health information organizations
Educational programs	Master's of Science in translational research	Master's of Science in clinical investigation	certificate and master's degree in comparative effectiveness; Web-based CME programs	Masters of Public Health; community education programs

experience and skills needed to participate in this area of translational research, either as a member of a research team or as a team leader/principal investigator.

Didactic Training Needs Continued Evolution of Coursework in the Theory and Methods of Translation

An expert working group of the CTSAs initially identified 14 core thematic areas with which to organize the core competencies in clinical and translational science (table 3)

[11]. Within these core thematic areas, over 100 specific core competencies were recommended across the entire spectrum of translational research. These core competencies provide guidance to institutions that are developing curricula. Moreover, for those institutions with longstanding curricula related to translational science, the core competencies can be mapped to existing courses, seminars, etc., as an inventory of didactic resources. The core competencies could also be used as an evaluation tool to monitor progress in upgrading didactic programs. Finally, the core curriculum developed by the CTSA institutions encourage engagement and assistance from specific disciplines (e.g. biostatistics, bioinformatics, etc.) to define coursework offered in these areas.

The National CTSA Educational Resource Program has undertaken the task to identify, catalog, and assess training modules supplemental to the core curriculum from among the 60 CTSA institutions; it has also included the large number of NIH courses archived for online use. Over 100 courses, 52 of them from NIH, have been identified and placed on a CTSA website [12].

The University of Iowa CTSA program has established a virtual university as an online repository of courses that can be shared by CTSA institutions, as well as with the national and worldwide research communities; access is through a secure internet portal [13]. The virtual university houses both core courses fulfilling the requirements of the core curriculum (e.g. biostatistics, epidemiology), as well as courses more customized for in-depth training. Educational resources are available in a variety of forms, including text (PDF), slides, podcast, and videocast.

The dynamic evolution of the translational research field is evidenced by the identification of new topical areas for consideration in the core curriculum. For example, comparative effectiveness research, which was not mentioned in the original funding announcement for the CTSA program, expanded rapidly in response to increasing interest shown by the NIH and the Obama administration, as demonstrated by commission of an Institute of Medicine report on national priorities for comparative effectiveness research (released in 2009) [14] and availability of significantly increased funding for comparative effectiveness research from the US government (Affordable Care Act) [15]. The establishment of the National Center for Advancing Translational Sciences further emphasizes the innovative and collaborative training programs needed to speed the delivery of new drugs, diagnostics, and medical devices to patients (http://www.ncats.nih.gov/index.html).

Clinical/Translational Research Training Should Be Competency-Based to Impart Practice-Based Skills

The core curriculum outlined in table 3 includes several thematic areas which are skill-oriented. For many knowledge-based learning objectives, the trainees' completion of a didactic course with an adequate evaluation/grade may suffice to assure that the competencies have been attained. However, information in a number of core thematic areas is not easily conveyed through didactic coursework. Examples include research implementation, responsible conduct of research, scientific

Table 3. The CTSA core curriculum: core thematic areas [11]

Clinical and translational research questions
Literature critique
Study design
Research implementation
Sources of error
Statistical approaches
Biomedical informatics
Responsible conduct of research
Scientific communication
Cultural diversity
Translational teamwork
Leadership
Cross-disciplinary training
Community engagement

communication, translational teamwork, leadership, and cross-disciplinary training. The University of Pittsburgh has been developing assessment methods for competencies not taught in didactic courses. Opportunities to assess skills, such as hypothesis formation, may occur during the course of training and during a grant-writing course. The requirement of scientific communication might be fulfilled by assessments of presentations at local and national research symposia or publications of abstracts and manuscripts. Other topics, such as 'responsible conduct of research' or 'leadership', might be addressed with case studies, role playing, or other simulations. Junior investigators need to receive training in those skills that are critical to their survival in the research environment; these include teaching, manuscript writing, public speaking, advocacy, grant application writing, team science, and leadership.

Effective Clinical and Translational Research Involves a Multidisciplinary Team of Co-Mentors

As depicted in figure 1, the multidisciplinary translational research team may include mentors from the basic, translational, and clinical sciences who are specifically chosen for their scientific discipline or methodological expertise. Table 2 illustrates multiple disciplines operating within the four domains across the translational research spectrum. This team-mentoring approach has a number of additional advantages, such as providing a broader range of role models for the trainees, additional opportunities for participation in diverse educational activities (e.g. seminars and journal clubs), and less reliance on one mentor to fulfill all the trainees' educational needs in areas such as preparing manuscripts and/or grants, accessing research technologies, and ensuring funding throughout the training period. The team-mentoring approach also provides options for the trainee if the primary mentor were to change, relocate, or retire.

The requirement to have multidisciplinary faculty 'teams' serve on doctoral thesis committees, as postdoctoral fellow advisors, or on career development mentoring committees, has resulted in the formation of multidisciplinary translational research teams. Departments and centers within research universities are frequently silos in which there is little involvement of faculty outside of one's own department or center. While many members of faculty may hesitate to take time away from their own laboratory to participate in a faculty-driven research collaboration, those same faculty members will often accept a trainee's request to join a student's research advisory committee, which in essence is a multidisciplinary translational research team. After the training has been completed, the team frequently continues on with mutually beneficial research collaborations.

Development of High-Quality Research Mentors Is Crucial to Expanding Translational Research

The development of mentors has been emphasized from the onset of the CTSA program. Regardless of the success of team approaches to mentoring new investigators, the need remains for one or two established faculty members to take responsibility for the career development of a trainee [16]. CTSA programs frequently scan their faculty ranks for mentors, often defined as those with established research funding and a track record of training students, fellows, and junior faculty. However, faculty with these qualifications may not always have optimal skills in mentoring. Two approaches to improving the quality and quantity of available mentors have been educational programs for current or prospective mentors and the development of mentoring tools and procedures to increase the quality and consistency of the mentoring activities.

A number of CTSA programs have initiated faculty development programs to enhance mentoring practices among faculty members who are training translational scientists. A mentor workgroup within the National CTSA Consortium has sought to identify best practices of mentoring within the CTSA institutions. For example, the University of Wisconsin CTSA program has developed the Entering Mentoring curriculum [Fleming M: Entering Mentoring, 2011, personal communication], which focuses on mentoring processes, including six competencies: (1) effective communication, (2) establishing and aligning mentor and mentee expectations, (3) assessing mentees' understanding of scientific research, (4) addressing diversity within mentoring relationships, (5) fostering mentees' independence, and (6) promoting mentees' professional career development. This curriculum has been evaluated at 16 institutions (most of them CTSAs) with 283 mentor-mentee pairs randomized to an 8-hour case-based training with small group discussion or to no intervention. A mentoring composite score was used to evaluate the intervention versus control group mentoring competency and practices, with results endorsing the use of structured training to improve mentoring skills. The training materials are available on a CTSA website [13].

Other institutions have set up more formal processes and tools to support high-quality mentoring. A tool increasingly required by NIH-funded training grants is the Research Career Development Plan or 'mentoring contract' in which the mentor and mentee use a template to create a mutually agreed upon training plan [16]. This written document facilitates communication and mentor/mentee expectations. Frequently, the RCDP includes an explicit description of the following: (1) the mentee's goals and learning objectives for the period of training; (2) the didactic coursework or degree program that supports those goals; (3) the mentored research project, including the multidisciplinary co-mentors that may be involved and a plan for interaction between the mentee and mentors; (4) a plan for career development activities (e.g. participating in or attending seminars, workshops, national conferences, summer courses), conducting teaching activities, and writing papers for publication and grant applications. The RCDP can then be used by the mentor/mentee dyad for annual review to assure the training program is making progress toward the mentees' career goals and learning objectives.

The RCDP also provides opportunities for mentor development. At the University of Pittsburgh, a structured program trains mentees on mentoring skills with the philosophy that the skills learned for their mentor-mentee dyads will also be learned by their mentors. At the University of Rochester, senior faculty members on a mentor development committee meet with all CTSA-supported trainees one-on-one in the first half of the academic year and with the mentor-mentee dyad in the second half of the academic year, to review progress with the RCDP. This provides an opportunity for monitoring and feedback to identify dysfunctional dyads and either redirect them or identify a new mentor for the trainee. Both approaches provide opportunities to improve the mentor's skills, including further participation in mentoring workshops and seminars throughout the year.

Clinical and Translational Research Educational Programs Should Be Coordinated and Promoted at the Institutional Level

In the USA, academic programs leading to master's and doctoral degrees have been based within departments that are under the purview of schools and colleges of the universities. Such administrative structures may serve as barriers to maximizing the interdisciplinary nature of training. Whereas didactic courses providing the core competencies may be located in specific departments or programs, the oversight of trainees' use of courses, mentors, seminars, etc., may benefit from a broader perspective. This may serve to reduce redundancy of course and seminar offerings, while enhancing the diversity of course participants and encouraging involvement of multidisciplinary faculty on research teams. For example, at the onset of their CTSA program, the University of Pittsburgh established the Institute for Clinical Research Education to address these issues. All clinical-translational science training programs were brought under one umbrella organization. A mentoring network of more than 400 faculty members was developed to provide mentors to students and trainees from virtually every department and division in their schools of health sciences, as well as other schools such as engineering, computer sciences, social work, etc.

Continuing Challenges and Opportunities

The seven premises provide guidance to the organization and implementation of translational research training programs. The CTSA program has been active and growing for over 6 years, but has only begun to address the national shortage of translational scientists. From a long list of issues facing translational research training, a number have received special attention due to recognition of their continuing importance. A major concern has been the final step in the career development, namely that of receipt and renewal of the individual research grants that provide a long-term source of funding for a research program [17]. This includes the need for the NIH review process to recognize this new breed of translational researchers, who are multidisciplinary and who are developing research proposals that frequently span several NIH Institutes' priorities, and may include both preclinical (e.g. cell and animal studies) and clinical studies in the same proposal. The CTSA program and its member institutions have initiated specific programs to minimize the dropout of translational investigators at this later stage of career development.

Another concern is the diversity of the US translational research workforce. The proportion of women and minority investigators enrolled in professional and graduate programs has continued to increase, but most senior faculty members in US universities continue to be nonminority males. The poor retention rate of women and minorities must be addressed. The University of Rochester has received a Pathfinder Grant Award from the NIH to assess the effect of different methods on retention of these new investigators in research careers. According to the evaluation protocol, 152 mentor/mentee dyads, which include a woman or underrepresented postdoctoral fellow or junior faculty member and his/her mentor, have been randomized to one of four groups: (1) a peer-mentoring program, (2) a behaviorally based mentoring program, (3) both programs, or (4) none of the above (i.e. controls) [Lewis V, personal communication].

A third area of interest is the education and training of investigators for team science, including the development of leadership skills in the context of a research team. However, appointment and promotion rules frequently do not recognize all of the faculty members who are essential to the function of research teams (e.g. those engaged in biostatistics, bioinformatics, basic science core laboratories). The paradox then is to develop the practice of team effort in training, but expect that the research career of the graduate may involve team research that has limited recognition of the individual for his/her effort.

Finally, the need for translational researchers either to work in industry or to perform research studies sponsored by industry continues to grow [18], despite marked shifts in private sector investment in basic science. However, curricula frequently do not expose the trainee to regulatory science or to the regulatory issues in drug or device development. The Public-Private-Partnership Key Function Group of the National CTSA Program has worked to develop competencies in drug and device development and technology transfer. Nonetheless, graduates from academic programs continue to require extensive training by industry, raising the question about whether some of this training should be incorporated into their curriculum and career development components of their translational research training programs [19].

References

1 National Institutes of Health: Setting Research Priorities at the National Institutes of Health. September 1997. http://www.nichd.nih.gov/research/planning/.

2 Nathan DG, Varmus HE: The National Institutes of Health and Clinical Research: a progress report. Nat Med 2000;6:1201–1204.

3 Lenfant C: Shattuck lecture: clinical research to clinical practice – lost in translation? N Engl J Med 2003;349:868–874.

4 Schwartz K, Vilquin JT: Building the translational highway: toward new partnerships between academia and the private sector. Nat Med 2003;9:493–495.

5 Sung NS, Crowley WF Jr, Genel M, et al: Central challenges facing the clinical research enterprise. JAMA 2003;289:1278–1287.

6 Nathan DG, Wilson JD: Clinical research and the NIH – a report card. N Engl J Med 2003;349:1860–1865.

7 Zerhouni EA: Translational and clinical science: time for a new vision. N Engl J Med 2005;353:1621–1623.

8 Collins F: Reengineering translational science: the time is right. Sci Transl Med 2011;3:90cm17.

9 Colwell R: Report of the Committee on Enhancing the Master's Degree in the Natural Sciences. Washington, National Academy of Sciences, 2007.

10 Pearson TA, Fogg TT, Bennett N, Kieburtz K, Kitzman H, Moxley R, Puzas E: Building capacity across the spectrum of research translation: centers of excellence within the Rochester Clinical and Translational Science Institute. Clin Transl Sci 2010;3:272–274.

11 CTSA Core Competencies. https://www.ctsacentral.org/core-competencies-clinical-and-translational-research.

12 National CTSA Educational Resource Program. https://research.urmc.rochester.edu/ncerp/search/.

13 CTSA Virtual University: http://www.icts.uiowa.edu/content/virtual-university.

14 Institute of Medicine: Initial National Priorities for Comparative Effectiveness Research. Washington, The National Academies Press, 2009.

15 Lauer MS, Collins FS: Using science to improve the nation's health system: NIH's commitment to comparative effectiveness research. JAMA 2010;303:2182–2183.

16 Johnson WB: On Being a Mentor. A Guide for Higher Education Faculty. New York, Psychology Press, 2007.

17 Kotchen TA, Lindquist T, Malik K, Ehrenfeld E: NIH peer review of grant applications for clinical research. JAMA 2004;291:836–843.

18 Stevens AJ, Jensen JJ, Wyller K, Kilgore PC, Chattergee S, Rohrbaugh ML: The role of public sector research in the discovery of drugs and vaccines. N Engl J Med 2011;364:535–541.

19 Biomedical Research Workforce Working Group Draft Report. National Institutes of Health. June 14, 2012. acd.od.nih.gov/bmw_report.pdf.

Thomas A. Pearson, MD, MPH, PhD
Rochester Clinical and Translational Science Institute, University of Rochester
265 Crittenden Boulevard, CU 420708
Rochester, NY 14642 (USA)
E-Mail Thomas_Pearson@urmc.rochester.edu

Alving B, Dai K, Chan SHH (eds): Translational Medicine – What, Why and How: An International Perspective.
Transl Res Biomed. Basel, Karger, 2013, vol 3, pp 47–55 (DOI: 10.1159/000343019)

Training Translational Investigators in China

Tim Zhanxiang Shi[a] · Kerong Dai[b]

[a]Global MD Organization Network Corp (GlobalMD), Catonsville, Md., USA; [b]Clinical Translational Center of Stem Cell and Regenerative Medicine, Shanghai Jiao Tong University School of Medicine, Shanghai, PR China

Abstract

As multinational clinical drug trials have been increasingly conducted in China, while more challenges are in ensuring the quality of research practice and protections of human research subjects. We have identified a specific need to introduce the standards and principles of clinical and translational research based upon the core training components of the National Institutes of Health (NIH) and the NIH founded CTSA institutions. The initial outcomes of training clinical and translational researchers in China were not only advanced the concepts, but also the expertises were able to be adapted, applied and shared within both the Chinese and the US environments.

China has been undergoing rapid economic growth over the past few decades. Subsequently, the national initiatives on scientific and technology development, specifically the National 11th Five-Year Plan (from year 2006 to 2010) and the National 12th Five-Year Plan (from year 2011 to 2015), have included significant investment in the infrastructure of bioscience and medical research [1]. Meanwhile, innovative drug research and translational research practices, along with evidenced-based policy making, have been highlighted and supported by the government [1]. Although it is too early to tell the benefits of clinical research for the 1.3 billion people of China, more clinical research facilities affiliated with local hospitals and academic institutions have been established to function as designated sites for clinical trials that will include, among other studies, testing of the safety and efficacy of new agents designed by industry sponsors. The expected outcome of high-quality translational and clinical research in drug development would be the more rapid availability and accessibility of standardized, efficacious products for the Chinese population.

Multinational clinical research and drug trials have increased substantially in China in recent years; moreover, increasing amounts of data collected in China and other developing countries are being submitted to the United States Food and Drug Administration (FDA) as part of industry sponsors' multisite trials and postmarketing surveys [2].

The globalization of clinical research has raised significant challenges in ensuring the application of good research principles and practices and the maintenance of data quality and protection for human research subjects [3–6].

Since the majority of clinical research studies and clinical trials are conducted by clinicians, nurses, and health care practitioners, the quality of their professional training is just as critical as the proper management of trial data. Medical education worldwide, however, remains oriented towards health care, i.e. managing patient treatment [7]. Little time is given to teaching clinical research concepts and practices. Furthermore, opportunities for formal courses and certification in clinical research are often limited in developing countries, such as in China. Without continued development of recognized international standards, global sharing and cooperation will be more difficult [6].

We have identified a specific need for standardization in training, which could facilitate communication and collaboration among the key stakeholders, including academia, health care services, and bioindustry, within and across countries with diverse social and economic environments.

Training Courses in China

Since Chinese health care practitioners at local hospitals and investigators at research institutes did not have easy access to advanced knowledge and systematic training opportunities in the international standards of clinical research, in 2008 we began to introduce a number of training courses into China (table 1). Initially, we selected the US National Institutes of Health Clinical Center (NIH CC) course on the Principles and Practice of Clinical Research (PPCR), which is a NIH core training component [8], due to its professional acclaim and international validation (over 20,000 professionals certified globally since 1995). Another training component was developed based upon the NIH Clinical and Translational Sciences Awards (CTSA) model, focusing on effective governance and leadership of clinical and translational research in the unique academic and health care environments in China. Courses developed from these two sources were synthesized into a 3- or 5-day series of intensive classroom lectures and panel discussions. As additional needs for training in clinical trial design and human research subjects protection were identified among Chinese translational investigators, courses in Principles of Clinical Pharmacology (also a 5-day course), and Bioethics and Institutional Review Board (2-day course) designed by the NIH CC were added to

Table 1. Examples of training courses conducted in China

a Course 1: Principles and Practice of Clinical Research

Day 1
Introduction and program overview: history of clinical research and NIH clinical center
Clinical research project design and guidelines: an introduction to biostatistics: randomization, hypothesis testing, and sample size

Day 2
Institutional review boards and integrity in clinical research and ethical issues
Sample institutional review board

Day 3
Data and safety monitoring in clinical research
Interactive input and discussion on protocols:
1. Liver cancer-gene therapy clinical project
2. Repair of peripheral nerve defect with human acellular nerve allograft: a multicenter evaluation

Day 4
Using secondary data in statistical analysis with illustrative examples unanticipated risk in clinical research
Interactive input and discussion on protocols:
1. Clinical research proposal: randomized control trial in locally advanced rectal cancer with neoadjuvant radiochemotherapy: long-term vs. short-term
2. A prospective multicenter surgical trial: the effect of ischemia/reperfusion injury elicited by the Pringle maneuver on the tumor recurrence of hepatocellular carcinoma

Day 5
NIH clinical research collaboration and NIH grant application
NIH instructors Q&A, NIH PPCR training evaluation
Final assessment and test

b Course 2: Clinical and Translational Research: Planning and Professional Management

Day 1
Introduction to the formation and progress of clinical and translational science activities in the USA
Clinical and translational research centers (CTRCs): defining the goals, building the leadership team, and developing the governance structure; role of the external advisory boards
Infrastructure requirements to conduct clinical trials at CTRCs: critical needs for trained research staff, project managers, biostatisticians, informatics tools
Integrating bioethics into clinical research

Day 2
Regulatory components in clinical and translational research practice
Building a CTRC consortium platform: informatics, data sharing and management of clinical trials in specific disease areas
Developing effective and efficient core facilities in CTRCs: discussion of administration and finances
Development of CTRC training programs to establish interdisciplinary, team-based approaches to clinical and translational research
Utilization and financial management of pilot projects to stimulate collaborations within and among CTRCs

Table 1. continued

Day 3 Establishing bioindustrial and academic collaborative partnerships and consortia (to include discussion of how to prevent conflicts of interest) Evaluating the effectiveness of the CTRC: designing quantitative and qualitative measures to assess in real-time the value of the center Panel discussions and translational center case studies Review: China National 12-5 Scientific Development Plan: supporting biomedical translational research Course summary and assessment: Q&A and discussion

c Course 3: Principles of Clinical Pharmacology

Day 1 Introduction to Principles of Clinical Pharmacology Drug metabolism Chemical assay of drugs and drug metabolites Clinical pharmacokinetics Questions and discussion
Day 2 Compartmental analysis of drug distribution Drug absorption and bioavailability Effects of renal disease on pharmacokinetics Pharmacokinetics in patients requiring hemodialysis Questions and discussion
Day 3 Pharmacogenetics Effects of liver disease on pharmacokinetics Population pharmacokinetics Physiological and laboratory markers of drug effect Questions and discussion
Day 4 Design of clinical drug development programs Disease progression models and clinical trial simulation Role of the FDA in guiding drug development Questions and discussion
Day 5 Course summary and assessment: Q&A and discussion

d Course 4: Bioethics and Introduction of Institutional Review Board (IRB) Practice

Day 1 Computer-based training for NIH IRB members: introduction, using the NIH IRB review standards, continuing IRB review, IRB minutes, and other issues

Table 1. continued

Day 2
NIH IRB new member orientation: institutional review boards, standard care, nonstandard care and research, case studies and discussion
'Cutting edge consent'
IRB topics:
1. Vulnerable populations
2. Waivers of informed consent
3. Waivers of documentation of consent
4. Embryonic stem cells
5. Differences in FDA and OHRP rules
6. Payments to subjects
7. Research in developing nations
8. Emergency research
9. Therapeutic misconception
10. Genetics research
Day 3
Introduction to research ethics: eight ethical principles; informed consent research
Risks and benefits: IRB requirement in the USA vs. EU GCP directive
Course summary and assessment: Q&A and discussion

complement the training program. The courses were taught by instructors from either the NIH CC or faculty from academic institutions that had received a NIH CTSA award. Thus, faculty were drawn from the CTSA at Yale University (Yale Center for Clinical Investigation), Vanderbilt University (Vanderbilt Institute for Clinical and Translational Research), Southern California University (Southern California Clinical and Translational Science Institute), and the University of Florida (University of Florida Clinical and Translational Science Institute). The faculty from these different institutions have provided Chinese physicians and clinical researchers with information about a wide variety of resources and shared their hands-on experiences, as well as lessons learned, in the development of their clinical and translational science centers.

The training provides necessary knowledge and information in a prospective systematic fashion, in contrast to information obtained through investigator experience. Furthermore, the training improves clinicians' research skills, helping to strengthen the clinical research-oriented environment in local hospitals. As such, the availability of multidisciplinary expertise has become a critical cornerstone for translational investigators in the local environment of healthcare and research.

A number of different teaching methods were used, including on-site lectures in China by NIH faculty, Web-based courses through the website www.ChinaGlobalMD.cn, and scientific workshops or conferences on specific topics. Additionally, textbooks and handouts from lectures were translated into Chinese to be used for personal study by trainees, and changes to the local problem-based case studies were made as needed for the different professional levels of trainees. A combination of training methods has been found to be most effective; courses now include lectures, online videos, and group study sessions on specific issues, as well as one-on-one mentoring.

As of July 2012, approximately 4,000 clinicians, basic researchers, and nurses from approximately 500 hospitals and research institutions across China have participated in the various clinical and translational research training courses organized by GlobalMD, NIH CC, and leading hospitals in China.

Based upon the data from trainee registration forms, about 48% of participants were clinicians, 28% were project assistants, and the remaining 24% were administrative staff, residents, or MD or PhD candidates. The majority (76%) of participants have been directly involved in clinical and translational research and/or drug trials in their daily practice. Their motivations for participation were principally knowledge acquisition, career development, and promotion. For both current students and recent graduates, the potential to have a career in clinical and translational research is also a motivating factor.

Evaluation of Training

We have evaluated the effectiveness of each of the courses through conducting surveys to be completed by attendees in an anonymous fashion. For instance, at the PPCR course (fig. 1), trainees took pre- and posttraining tests to evaluate the impact of the training on their grasp of clinical research skills. Questionnaires given to participants clinical research environments before and immediately after each training course showed that their conceptual and practical knowledge of the subject matter improved significantly as a result of the courses.

A postcourse assessment, using a four-degree Likert scale (which is a psychometric scale used in questionnaires, with ratings of excellent, good, fair, and not good), showed that content preparation was rated as excellent (90%), as was the quality of lectures (90%). Case discussions were rated only good (67%), likely because these cases were based on US clinical research environments and differed from those of Chinese local hospitals, particularly regarding informed consent and protection issues for human subjects. Eighty-seven percent of survey respondents believed that the course provided an opportunity to compare and understand the differences in the principles and practices of clinical research between the USA and China, while 94% agreed that the clinical research methods they had learned were applicable to clinical trial practice in China.

To assess the long-term impact of training on the clinical research practices of attendees, 100 of the original participants were randomly chosen to receive a follow-up survey after 18 months. Participants were questioned on whether the course led to improved understanding of the concepts and practice of clinical research, if it motivated greater interest in clinical research practice, if the PPCR was understood and

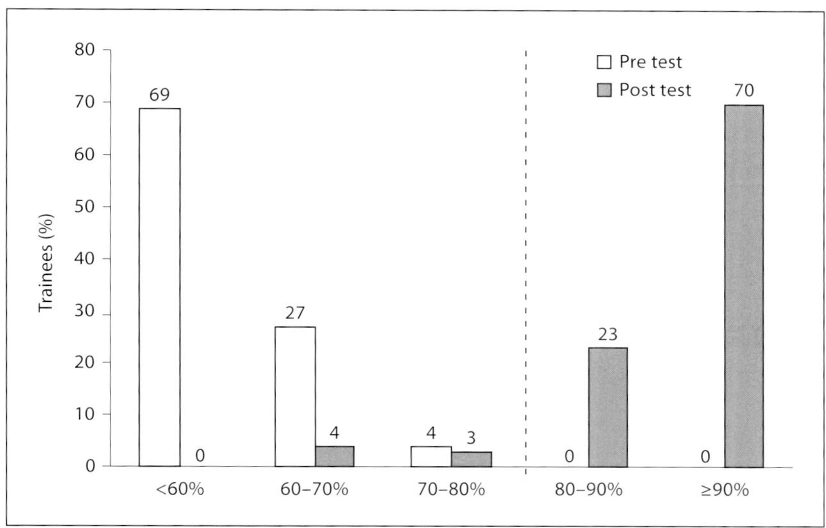

Fig. 1. Pre- and posttraining assessment of the NIH PPCR courses: percentage score on test.

adopted, and if the methods they learned were actually being applied. Participants were queried about knowledge of data safety monitoring, awareness of adverse events reporting, understanding of postmarket monitoring, and benefits of the course for their disease specialty clinical research, success of grant applications written, and number of scientific publications. The response rate for this survey was 100%, and the results again were outstandingly positive, with all aspects of the training judged to be extremely or very useful.

A separate survey was also conducted anonymously for the course 'Clinical and Translational Research: Planning and Professional Management' on the last day of training activities after a full 3-day program (table 1). The satisfaction rating for the training activities was 97%, the format of lectures and presentations garnered an approval rate of 93%, and the format of panel discussions was given the highest satisfaction rate of 100%. In addition, the trainees evaluated the potential benefit for clinical and translational research practice in their own work environment. The majority of trainees (89%) considered the training program to be useful in better understanding the concept of translational science, and 79% indicated that it could advance their translational research practice. However, 76% of the trainees indicated that it seemed unlikely that clinical therapeutics and/or basic laboratory research practice would directly benefit immediately from the training.

Trainees expressed great interest in the following areas of course content: case studies of successful translational research projects, the hospital's role in advancing translational research, pathways in teamwork and collaboration, resources of both academic and industrial collaboration on translational research, and lastly and most importantly, the issue of intellectual property and scientific publication authorship in translational research practice.

Although China has not yet clearly defined a roadmap of translational science development at the national level, it is clear that the Chinese model of translational research practice would be a different approach than the NIH's roadmap for advancing translational sciences. It would be more focused on the improvement of health care and disease prevention at the community level.

Conclusion

During the past 4 years, clinical and translational research principles and practice have been introduced to professionals across multiple specialties in China. Evaluations of our courses by the attendees have helped us to adjust the content and the methods of teaching to ensure maximum benefit for the participants. Clear advances in clinical research education and practice have been achieved, and the possibility for standardization and true globalization of translational medicine is getting stronger. This experience has illustrated that the principles of clinical and translational research and valuable expertise developed in the USA can be adapted, applied, and shared within both the Chinese and the US environments.

As a next step to improve the quality of clinical and translational research, a standardized conception of clinical research and protection for human research subjects will be further supported by developing a virtual network among the Sino-American medical professionals. This network will provide a platform for effective and meaningful communication and facilitate the translation of knowledge and expertise gained from clinical research into the practice of community physicians.

In order to provide multidisciplinary resources for local physicians and researchers who are learning about clinical and translational research, we collaborated with NIH-funded academic health centers with the CTSA awards to adapt their education and training programs to train the next generation of translational investigators in China. We have learned from our course evaluations that for the Chinese investigators, the most effective training in bioethics and protection for human research subjects would be based on examples of specific situations occurring in China in clinical trials, research laboratories, and clinical wards.

Multidisciplinary research teams and standardization of good clinical practices and regulation are required to catalyze translational science development in China. An effective training program is a fundamental element of the infrastructure of translational science in China, and likely throughout other countries as well. Education and training of the next generation of clinical and translational investigators can and should be one of the primary methods to minimize differences in clinical and translational research standards and practices as the development of clinical and translational research continues around the world.

References

1 Chen Z: National plan on translational medicine in China: promoting health care reform and improving people's health; in Selected Presentations from the 2011 Sino-American Symposium on Clinical and Translational Medicine. Washington, Science AAAS, 2011, pp 7–8.

2 FDA Cardiovascular and Renal Drugs Advisory Committee. 2007 Meeting Documents. http://www.fda.gov/ohrms/dockets/ac/07/transcripts/2007-4327t-02-part2.pdf (accessed Jan. 30, 2009).

3 Glickman SW, McHutchison JG, Peterson ED, Cairns CB, Harrington RA, Califf RM, Schulman KA: Ethical and scientific implications of the globalization of clinical research. N Engl J Med 2009; 360:816–823.

4 Thiers FA, Sinskey AJ, Berndt ER: Trends in the globalization of clinical trials. Nat Rev Drug Discov 2008;7:13–14.

5 Rowland C: Clinical trials seen shifting overseas. Int J Health Serv 2004;34:555–556.

6 Califf RM: Simple principles of clinical trials remain powerful. JAMA 2009;293:489–491.

7 Annas GJ: Globalized clinical trials and informed consent. N Engl J Med 2009;360:2050–2053.

8 Gallin JI, Ognibene FP (eds): Principles and Practice of Clinical Research, ed 2. Burlington, Elsevier, 2007.

Tim Zhanxiang Shi, MD, PhD
5632 Baltimore National Pike, Suite 202
Catonsville, Md. 21228 (USA)
E-Mail Tim@GlobalMD.org

Alving B, Dai K, Chan SHH (eds): Translational Medicine – What, Why and How: An International Perspective.
Transl Res Biomed. Basel, Karger, 2013, vol 3, pp 56–62 (DOI: 10.1159/000343027)

Translational Medicine in Taiwan: The Role of Private Foundation Funding

Samuel H.H. Chan

Center for Translational Research in Biomedical Sciences, Kaohsiung Chang Gung Memorial Hospital, Kaohsiung, Taiwan

Abstract

A new model for translational medicine has been created at the Center for Translational Research in Biomedical Sciences (CTRBS) at Kaohsiung Chang Gung Memorial Hospital in the southern city of Kaohsiung in Taiwan. This chapter describes how the core value of a private enterprise in Taiwan has resulted in the very first translational medicine center in Taiwan, along with the strategies employed by Chang Gung Medical Foundation to use private funding to implement the philosophy and goal of the CTRBS in advancing biomedical research, especially translational medicine.

Over the past four decades, Taiwan, which has a current population of approximately 25 million people, has experienced a growing economy. Like many countries in the world, Taiwan provides governmental support for biomedical research through its various ministries. In the arena of translational medicine, one major development in recent years is the National Research Program for Biopharmaceuticals launched in 2011 [1]. The funding agencies for this program include the National Science Council, Ministry of Economic Affairs, Department of Health, and Atomic Energy Council, and the participating organizations include the Development Center for Biotechnology, National Health Research Institutes, and Center for Drug Evaluation. This program purportedly provides a platform for communications, interactions, and collaborations between government, academia and industry, so as to increase the willingness of enterprises to invest in the field of biomedicine.

Governmental funding for initiatives in translational medicine in Taiwan is complicated by three factors: (1) the term 'translational medicine' has not been clearly defined at a national level; (2) different ministries have diversified and sometimes conflicting missions; and (3) the nature of bureaucracy is such that policies may not be sustained, resulting in unstable long-term funding. However, private funding from the Chang Gung Medical Foundation (CGMF) in Taiwan has played a critical role in advancing biomedical research, especially translational medicine.

The CGMF was established in 1976 by Yung-Ching Wang, the founder of the very large and successful Taiwan-based Formosa Plastics Group. The mission of the foundation is to improve patient care through creating an integrated and comprehensive health care system, recruiting and training talented investigators, using the most advanced technologies for research and for patient care, and pioneering health care changes to maximize efficiency. With the objective 'to become a provider of health services and an advocate for public welfare based not on revenue prerequisites', CGMF has developed six medical centers that are strategically positioned to serve the populace of Taiwan. Medical research has always been an integral part of all these medical centers, with liberal funding from the foundation. In 2007, Y.C. Wang commissioned the development of the first translational medicine center in Taiwan, and in 2009 the Center for Translational Research in Biomedical Sciences (CTRBS) was inaugurated at the Kaohsiung Chang Gung Memorial Hospital, the second largest of the six medical centers.

The Philosophy and Goal of the Center for Translational Research in Biomedical Sciences

The governing philosophy of the CTRBS is twofold [2]. First, translational medicine is about solutions to health problems. The goal of medical research is the betterment of humankind, and the only difference between clinical and preclinical research is the study material. As such, translational medicine in effect transcends the boundaries between bench and bedside research, and is the synthesis of ideas, technologies, and research outcomes that are associated with a particular health theme. Second, translational medicine is about communication. The needs for resource discovery and collaboration tools by translational scientists are already recognized [3]. The word 'translation' is most commonly defined as the expression of words in another language. Its definition can be extended to encompass expression in simpler language and uncomplicated interpretation. Translational medicine is therefore the presentation of knowledge gathered in a fashion that is amenable to nonexperts. A molecular biologist is a nonexpert when it comes to interpretation of the clinical ramifications of acute myocardial infarction; a cardiologist is a nonexpert when it comes to interpretation of the intricacies of DNA repair. Communication in simple, understandable language will therefore bridge the intellectual gap between laboratory scientists and clinical practitioners. Under this philosophy, the goal of the CTRBS is to foster a congenial environment where MDs and PhDs can interact freely in terms of generating meaningful research problems and solutions without the worry of availability of facilities, know-how, funding, free time, or clinical significance. This ideal situation is advocated in a recent editorial on translational research [4].

Strategies of the Center for Translational Research in Biomedical Sciences

Provision of State-of-the-Art Facilities and Technical Support

One of the obstacles in contemporary medical research is the timely availability of appropriate facilities, and is usually a hindrance for both MDs and PhDs who wish to address a competitive health question. To overcome this obstacle, the CTRBS has, during the last 3 years, focused on purchasing, installing, and testing new state-of-the art equipment. In essence, facilities for physiological and biochemical experiments at molecular, cellular, tissue, and whole animal levels are currently available to complement modern clinical facilities in the medical center. The layout of these facilities is based on an open laboratory concept to maximize interactions between the users. In addition, an AAALAC International-accredited Center for Laboratory Animals provides excellent support for experimental animals, including genetically manipulated mice. Also available is a proteomics and genomics core that provides services for studies involving massive analysis at the protein, RNA, DNA, or microRNA level, including single-point mutation. A tissue bank ensures proper collection, preservation, and dispensing of clinical specimens in accordance to government regulations. Phase II refurbishment of additional laboratory space is currently underway, with procurement of more state-of-the-art equipment. To ensure proper handling of the facilities, managers are assigned to all major equipment, and users must complete training and receive certification before they are allowed to reserve instrument time. More than 55 training sessions have been carried out yearly for this purpose.

Compatible Clinical and Laboratory Facilities

To befit translational research, compatible clinical and laboratory facilities are installed in the CTRBS. For example, the presentation of images from the animal ultrasonograph (fig. 1) or animal MRI and their analysis programs are compatible to those used in the clinics at the medical center, making cross-interpretation between disease symptoms and phenotypes from animal models easily accessible.

Provision of Scientific and Clinical Advice

State-of-the-art facilities notwithstanding, the quality of medical research resides in employing the most appropriate and sensitive methodology in addressing scientific issues. In addition to their individual research endeavors, PhDs at the CTRBS routinely provide consultations to MDs on the best methods to answer a mechanistic question. Likewise, MDs provide advice to PhDs on the validity of, for example, an animal model based on whether the stipulated phenotypes are actually manifested by patients with the targeted disease. In the end, this is a win-win situation for both the clinical and laboratory scientists. The CTRBS also arranges regular meetings with two other members of the Chang Gung Group funded by the CGMF. They are the Chang Gung Biotechnology Corporation and Chang Gung Medical Technology Company; both offer a channel for the development of potential commercial products or diagnostic tools.

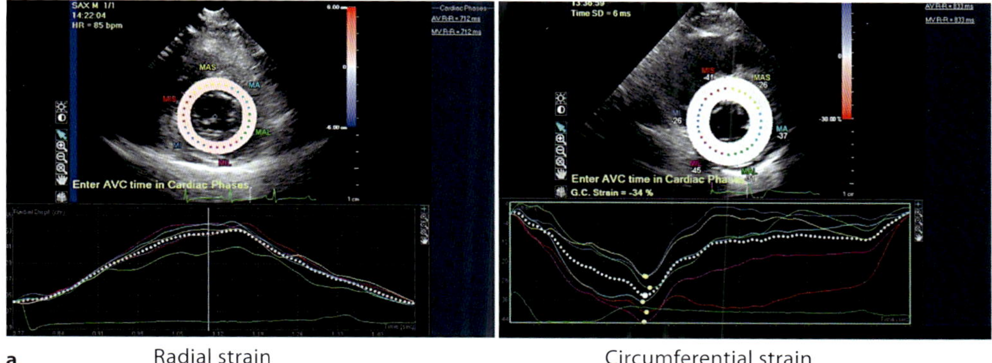

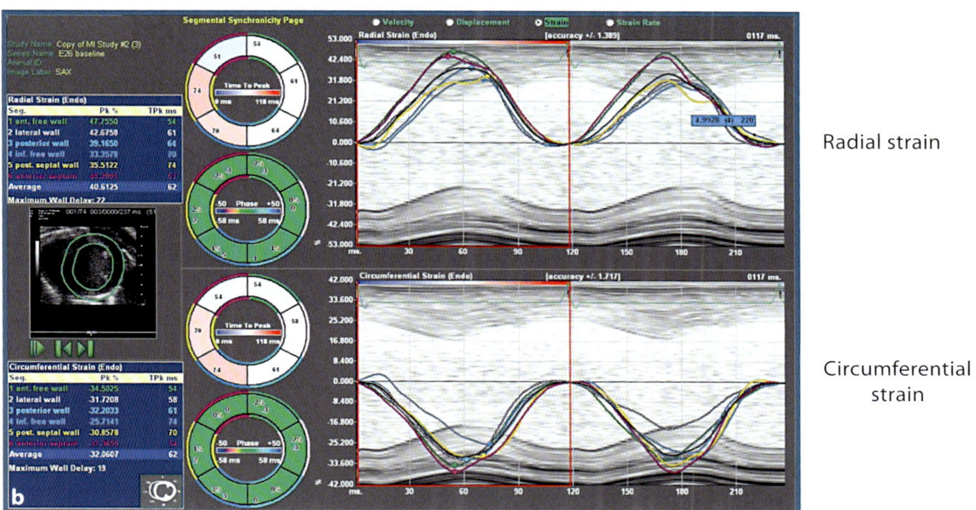

Fig. 1. Illustrative examples of compatible clinical and laboratory facilities. Strain images obtained from human (**a**) or mouse (**b**) ventricle showing comparable presentations of radial or circumferential strain measurements (images kindly provided by Sarah Chua, MD, Kaohsiung Chang Gung Memorial Hospital).

Provision of Financial Support

The principal concern on research projects at the CTRBS is whether the study is of health significance. Beyond that, funds for manpower and supplies are not an issue, and facilities are already available. However, members of the CTRBS must secure funding via formal application to the CGMF (see below). In effect, this practice is to make use of these procedures as a means to solidify the health problems to be addressed, and more often, to put the know-how of clinical and laboratory scientists together to provide comprehensive answers. In the spirit of translational medicine and within the boundaries of medical ethics on human investigations, program project grants are encouraged, in which association studies from clinical evaluations are complemented by mechanistic delineations in animals or cell models.

Allied Health Personnel
A very unique program at the Kaohsiung Chang Gung Memorial Hospital is that allied health personnel are also encouraged to participate in research activities. These research endeavors in general address issues that are immediately related to the professional activities of the nursing staff, pharmacists, radiographers, and medical technologists or therapists of various disciplines. We have learned that their enthusiasm in research has in many instances been the driving force for physicians working in the same unit to join them as a research partner.

Intramural Supports from the Chang Gung Medical Foundation

The CGMF provides generous financial support to all its six medical centers, including the CTRBS, through various intramural programs.

Research Grants
To encourage medical research, the Ministry of Health in Taiwan requires all medical centers to spend at least 3% of its gross annual income in research. For the CGMF, this translated to approximately USD 80 million in 2011, all of which was used for supporting research activities. One of these is in the form of research grants, which is administered independently by a research grants committee established in each medical center. There are three rounds of applications each year, and funding is for 1–3 years. The success rate is high because the basic philosophy is to provide support for the beginners to establish a research program, and for the more seasoned investigators to carry out pilot studies on more provocative projects. Nevertheless, despite this liberal support for research, all members of the Chang Gung family, including those in the CTRBS, are encouraged to seek external funding from government bodies. A matching fund amounting to 50% of the external grant is awarded as an enticement. In addition, publications and research grants form part of the assessment of performance to determine salary increase and promotion.

Training
Translational medicine relies on specialist knowledge provided by professionals who straddle research and therapy and possess both medical and scientific expertise [5]. Thus, several training programs are available to encourage clinicians to enter into medical research. MDs who have completed their residency training and have secured a staff position in any of the six medical centers can enroll in the Institute of Clinical Medicine of Chang Gung University to receive PhD training in biomedicine. They

can also do the same in a university abroad with full salary plus travelling expenses and a stipend for their tuition fees and subsistence. Alternatively, they can choose to receive 1–2 years of research training abroad, again with full salary plus travelling expenses and subsistence. The only requirement is that they must return to their posts and start a research program on completion of their degree or training.

Physician-Scientist Program
The foundation is fully aware that in the real world, there are two handicaps that prevent clinicians from being actively engaged in medical research: lack of free time and financial burden. This is particularly acute in Taiwan because a key determinant of the salary is clinical service. To overcome both handicaps, a physician-scientist program has been instituted in which a clinician agrees, with support from the department head, to devote 50% of his/her time to medical research. To compensate for the loss of income from clinical services, a stipend that amounts to 50% of the average salary of physicians of the same specialty and comparable seniority will be awarded. In addition, these physician-scientists are provided with a mentor to guide their professional growth. Recognizing the importance of mentoring skills, a training program fashioned after the University of California, San Francisco Clinical and Translational Science Institute Mentor Development Program [6, 7] is currently being developed.

Medical Ethics
The CGMF has a long history of emphasis on medical ethics in its clinical practice and research endeavors. Although each medical center is independent in terms of its daily operations, there is only one single institutional review board. This is to ensure uniform standards for research involving human subjects, and facilitate research programs that are carried out by staff from different medical centers. In addition, publications from members of the Chang Gung family are selected randomly every 3 months for scrutiny by the Research Ethics Committee for adherence to the spirit of the Institutional Review Board or Institutional Animal Care and Use Committee.

Conclusion

Under the mission of 'to serve, to educate, and to explore', medical research has always been an integral part of the CGMF. The CTRBS, which was established as a model of how translational medicine in a medical center should be carried out, epitomizes this core value of the foundation. The CTRBS model is now being adopted by the other medical centers in the Chang Gung family. Planning is currently underway for another center of integrated bedside-bench research at the Linkou campus, which houses the largest medical center of the Chang Gung group of hospitals. With sustainable financial commitments from the foundation, it is likely that the concept that translational medicine is for the betterment of humankind will take root.

References

1 National Research Program for Biopharmaceuticals. http://nrpb.sinica.edu.tw/en/home.

2 Chan SHH: Center for Translational Research in Biomedical Sciences, Chang Gung Memorial Hospital – Kaohsiung Medical Center, Taiwan; in Saunders S (ed): Selected Presentations from the 2011 Sino-American Symposium on Clinical and Translational Medicine. Washington, Science AAAS, 2011, p 23.

3 Bhavnani SK, Warden M, Zheng K, Hill M, Athey BD: Researchers' needs for resource discovery and collaboration tools: a qualitative investigation of translational scientists. J Med Internet Res 2012;14:e75.

4 Hobin JA, Galbraith RA: Engaging basic scientists in translational research. FASEB J 2012;26:2227–2230.

5 Wilson-Kovacs DM, Hauskeller C: The clinician-scientist: professional dynamics in clinical stem cell research. Sociol Health Illn 2012;34:497–512.

6 Feldman MD, Huang L, Guglielmo BJ, Jordan R, Kahn J, Creasman JM, Wiener-Kronish JP, Lee KA, Tehrani A, Yaffe K, Brown JS: Training the next generation of research mentors: The University of California, San Francisco, Clinical & Translational Science Institute Mentor Development Program. Clin Transl Sci 2009;2:216–221.

7 Feldman MD, Steinauer JE, Khalili M, Huang L, Kahn JS, Lee KA, Creasman J, Brown JS: A mentor development program for clinical translational science faculty leads to sustained, improved confidence in mentoring skills. Clin Transl Sci 2012;5:362–367.

Samuel H.H. Chan, PhD
Center for Translational Research in Biomedical Sciences, Kaohsiung Chang Gung Memorial Hospital
123, Ta Pei Road
Kaohsiung 83301, Taiwan
E-Mail shhchan@adm.cgmh.org.tw

Alving B, Dai K, Chan SHH (eds): Translational Medicine – What, Why and How: An International Perspective.
Transl Res Biomed. Basel, Karger, 2013, vol 3, pp 63–73 (DOI: 10.1159/000343154)

Utilizing Pilot Funding and Other Incentives to Stimulate Interdisciplinary Research

Scott C. Denne[a] · Tammy Sajdyk[a] · Christine A. Sorkness[b] · Marc K. Drezner[b] · Anantha Shekhar[a]

[a]Indiana Clinical and Translational Sciences Institute, Indiana University School of Medicine, Indianapolis, Ind., and [b]Institute for Clinical and Translational Research, University of Wisconsin-Madison, Madison, Wisc., USA

Abstract

At least 60 academic health centers throughout the USA are engaged in increasing the efficiency and quality of translating basic discovery into clinical trials and studies and into the community. Such efforts require teams of investigators from different disciplines who can address the complexities of translation. These institutions, which have received funding from the National Institutes of Health (NIH) through an initiative known as Clinical and Translational Science Awards (CTSAs), are allocating some of their resources to issue grants for pilot projects as a means of stimulating interdisciplinary research and supporting junior investigators in their career development. We discuss the ways in which pilot funding programs have been deployed at two different CTSA institutions: the Indiana University School of Medicine in Indianapolis and the University of Wisconsin-Madison. Both programs have found that pilot funds can be used not only to engage investigators, but also to influence their research interests, approaches, and collaboration choices. With a few years of continued implementation, sufficient faculty are engaged in the process to reach what may be a 'tipping point' for change in the institutional culture and environment. There are similarities and differences between the programs discussed in this chapter, yet each has experienced success in terms of dollars invested, investigators trained, and initiation of new research opportunities.

New biomedical discoveries at the basic science level are occurring at an extremely rapid pace, but the progression of these new ideas into the market as therapeutic products and the implementation of new findings into routine disease management are facing enormous delays [1]. Thus, to fully realize the benefits of the advances in basic biomedical research, we need to build new mechanisms which facilitate more rapid progression of discoveries. In order to accelerate the translation of emerging biomedical knowledge into well-implemented therapies, we need multidisciplinary research teams that can create ways of rapidly transporting new molecules, devices, technologies, or evidence-based interventions into testing, validating, and adopting into human health care practices, as well as make use of the evidence that arises from clinical trials and outcomes to inform new questions and approaches in basic science research [2]. Such a new paradigm requires 'translational research teams', comprised of members with a wide range of expertise ranging from biology to business. Such teams may involve collaborations among professionals from different backgrounds, such as basic scientists, clinicians (physicians, nurses, dentists, and veterinarians), health professionals (pharmacists, allied health), biomedical engineers, sociologists, epidemiologists, and others [3]. However, the majority of research funding is still focused on traditional research models, and there is little incentive (in fact, many disincentives) for scientists to conduct high-risk, collaborative research [4]. The recognition that novel funding approaches will be required to create team-based research was apparent to us as we began developing our applications for the Clinical and Translational Science Awards (CTSAs) sponsored by the National Institutes of Health (NIH) [5, 6]. In this chapter, we will discuss pilot funding programs, developed in the CTSAs at the University of Indiana Medical Center in Indianapolis and at the University of Wisconsin-Madison. Both were designed to encourage interdisciplinary research and to seed the creation of successful multidisciplinary research teams.

The process of translation from the laboratory to the community has been described in many different phases, with designated 'T' symbols for translation [7]. In this chapter we use 'T1' to designate research involving translation of laboratory-based knowledge into early studies in humans; T1 therefore includes a strong clinical component. 'T2' signifies research focused on translating results from clinical studies into practice and health decisions.

The Pilot Awards Program at the Indiana Clinical and Translational Science Institute

During our planning stage, we identified several critical obstacles for investigators interested in building multidisciplinary, translational research projects. These include: (1) deficiency of expert guidance and mentorship in multidisciplinary research, (2) lack of coordinated and easy access to resources or infrastructure, (3) insufficient protected time for experts to assist new investigators with preparative work, (4) complexity of multiple regulatory submissions, (5) lack of coordinated access to patient populations and health systems for recruitment of participants into clinical trials, and (6) a scarcity of readily accessible pilot funds to conduct such research.

To address these barriers, we created project development teams (PDTs) to provide in a structured fashion consultation and project management assistance. Each of these teams includes experienced investigators from diverse disciplines and a statistician, supplemented with a designated project manager and staff with expertise in regulatory affairs, intellectual property, contracts, and commercialization. A PDT is able not only to assist in developing the science project, but all such proposals are now enriched with project mapping, regulatory approval, and appropriate core resource requests. This approach has many advantages: (1) it increases the likelihood of investigators receiving input from many experts at once, (2) the experts themselves can interact with each other within the context of a PDT, (3) investigators can develop possible alternative approaches if necessary in consultation with experienced advisors, (4) necessary resources can be easily identified and rapidly accessed, (5) collaborations are facilitated between basic and clinical scientists, and (6) there is continual project management of the research to ensure the most efficient and beneficial process for investigators to reach their stated goals.

In 2008, we received CTSA funding and began to implement four PDTs: 'Preclinical PDT', a committee focused on translational studies involving animal and cellular models; 'Clinical Sciences PDT', a group focused on T1 and T2 studies in a wide variety of areas related to adult medicine; 'Translating Research into Practice PDT', a team dedicated to evaluating and implementing evidence-based practice and other health services research; and 'Pediatric PDT', a team devoted to research in children with an emphasis on interactions with basic scientists to design and implement bench-to-bedside studies. These teams serve as a 'one-stop shop' by providing investigators access to multidisciplinary research expertise, statistical and design support, pilot funds, core resources, project management, and other services. The members of these teams encompass a variety of disciplines, such as medicine, sociology, dentistry, biochemistry and molecular medicine, veterinary medicine, biomedical engineering, foods and nutrition, pediatrics, basic science, biostatistics, nursing, public health, behavioral studies, and regulatory support. PDT members are not simply review groups, but rather a team of expert individuals who discuss ideas and concepts with basic scientists and clinicians, including very junior researchers, and who assist them in fully developing high-quality, competitive clinical and translational research projects.

The Project Development Team Process

To assist investigators in initiating the Web-based PDT process, we created one entry point which is managed by a single individual, the translational sciences research officer. This design has been efficient and timely in guiding investigators to resources, while also creating an effective process for collecting metrics on the pilot funding program. The PDT process can be summarized in six steps detailed below. For a simple diagram of the process see figure 1.

Step 1: An investigator accesses IndianaCTSI.org and completes the brief PDT application. In the application, the primary investigator (PI) describes the type of assistance requested (e.g. funds for pilot data, statistical support, or assistance for reading a reviewer's comments from a recently submitted external grant), the research idea (anything from an initial idea to a fully developed protocol), and general information about his/her team of collaborators (including the PI).

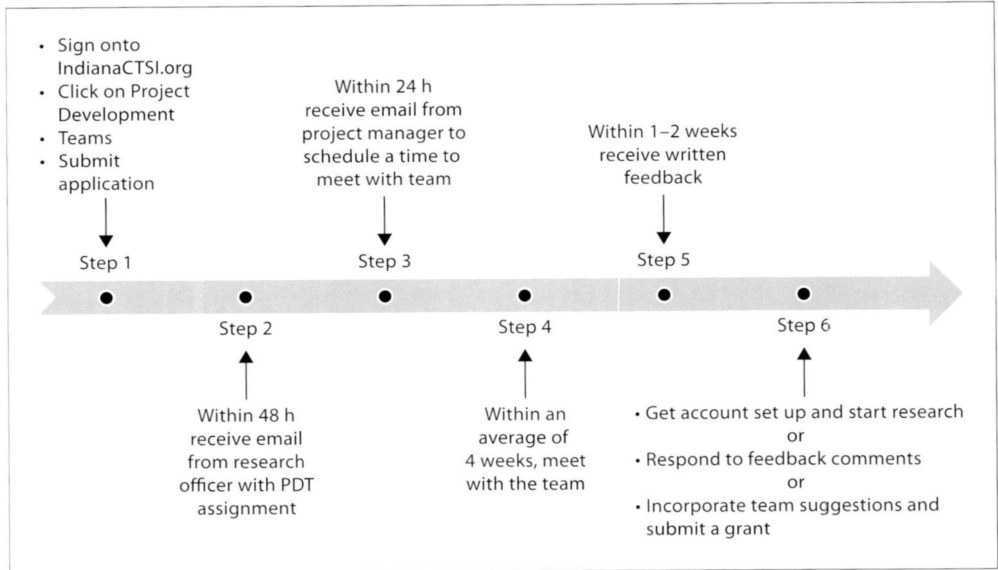

Fig. 1. A simple schematic of the Web-based PDT process.

Step 2: The Indiana Clinical and Translational Science Institute translational sciences research officer is immediately notified via email that an application has been submitted. It is reviewed and assigned to the appropriate team within 48 h. Once the project is assigned to a PDT, the PI is sent an email with the PTD assignment, as well as the name of the project manager for the team.

Step 3: The project manager from the PI's assigned PDT contacts the investigator within 24 h to arrange a time for meeting with the team.

Step 4: The PI and research colleagues meet with the PDT within 4 weeks (on average) and give a brief informal presentation of the project. During the meeting, which lasts about an hour, the PDT members and the research team have the opportunity to discuss the project and the needs of the investigators to move the study forward.

Step 5: The PI receives a feedback letter from the PDT within 1–2 weeks outlining the key points from the discussion and the overall outcome.

Step 6: The action required for this step depends on the outcome of the meeting. If pilot funding or other resources are approved, then an account is set up within 1–2 weeks. If the PI was given items to address before a funding decision can be made, the PI will prepare a written response to the team.

If the PI requested assistance on external grant reviews in preparing for a resubmission, the PI may incorporate the desired suggestions. If sufficient time is available prior to resubmission, the PI will send the revision or at least the specific aims page to the PDT for an additional review before resubmission to the external funding agency. A PI can request PDT assistance, meet with the team, and receive feedback on his/her

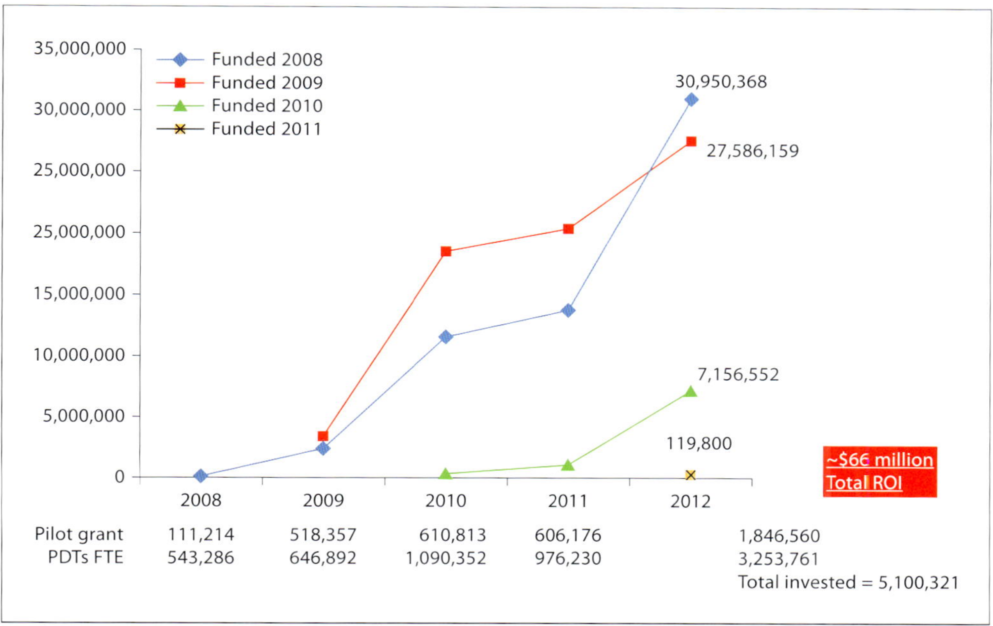

Fig. 2. A graph indicating the amount of extramural grant dollars generated each year by pilot grants awarded from May 2008 to February 2012 (cumulative total). PDTs provide to faculty the following: pilot grants (pilot grant dollars under the graph) as well as substantial ongoing assistance with expert reviews, mentoring, project management, and core resource support (cost of the PDT personnel is in terms of FTEs, which means full time equivalents or full-time staff). ROI = Return on investment. All values are in USD.

proposal in approximately 6 weeks. This provides a rapid and efficient process for the PI so that external funding deadlines can be met.

Metrics and Outcomes

After four full years of operation, we have found the PDT program to be a very successful component in working toward our mission. The success of this process is reflected in the overall number of projects that have come to the PDTs in 4 years (fig. 2). In addition to assisting a large number of investigators at all stages of career development, this program has resulted in a high return on investment dollars through external funding obtained from pilot grants within the first 3 years (fig. 2) and a robust number of new intellectual property filings. The intellectual property includes six licenses granted to industry, 18 disclosures filed, 22 patents (issued or pending), and eight start-up companies formed. The PDTs actively created 32 new collaborations between basic and clinical scientists and have successfully connected three universities and four campuses across the state of Indiana.

The PDT program has engaged faculty from 88 different departments across all institutions. PIs have reported that even if they did not receive actual dollars, they were still very enthusiastic about the other resources received. In year 3, all investigators who met with one of the PDTs but did not receive actual dollars were surveyed to assess the added value of the PDTs. Of the 84% of the PIs who responded, 79% said they would bring another project to a PDT, 64% said they made changes to their protocol based on the committee's input, and 53% indicated that the committee's input was helpful for submitting a grant for external funding. Such findings strongly suggest that the investigators who interact with the PDTs do not seek only financial support, but also find the program to be highly useful in facilitating their clinical and translational research projects.

The Pilot Awards Program at the University of Wisconsin-Madison Institute for Clinical and Translational Research

During our planning stage to develop an application for a CTSA, we established the goal of creating an environment in which research is a continuum of translating basic discoveries into professional practice, ultimately leading to practical improvements in health. Among the challenges identified were the absence of multidisciplinary research teams, which are essential to the performance of integrated translational research; the relative paucity of productive T2 research, an essential requisite for delivery of improvements in health care to the communities in Wisconsin; and the limited number of junior faculty successfully launching research careers. Therefore, we identified as a high priority the training of junior investigators; providing critical research resources necessary for T1 and T2 research; and fostering the development of multidisciplinary research teams, as well as novel research methodologies and technologies. The Institute for Clinical and Translational Research (ICTR) leadership instituted the Pilot Awards Program which (1) limited applicants for pilot awards primarily to junior investigators; (2) encouraged new or established junior-senior faculty collaborations to create an effective mentor-mentee relationship; (3) focused funding on research requiring interdisciplinary collaborations, enhancing the learning opportunities for the junior faculty; and (4) established the necessary support systems to foster pursuit of T1 and T2 research.

Essential to the success of the Pilot Awards Program was assuring an appropriate scope of the applicant pool, creating a novel review process for the applications, monitoring awardee progress and tracking award outcomes, and initiating/sustaining strategic process improvements to assure improved applications, support success of the projects, and facilitate rapid implementation of planned research programs.

Scope of the Applicant Pool

Requests for applications for funding for the Pilot Awards Program are released annually and are developed by program teams (T1 and T2 teams), in coordination with the ICTR administration, with input from the ICTR Board of Governors and key faculty partners. To model the applicant pool, ICTR leadership limited eligibility for a pilot award primarily to junior investigators and required that the applications reflect: (1) new junior-senior faculty collaborations; (2) development of novel research methodologies; (3) a focus on targeted research areas; and (4) creation of interdisciplinary research teams, including faculty from our partner institution, Marshfield Clinic (a large health care system in Wisconsin with two hospitals and 52 community care centers and an associated research and education foundation). The requests for applications have been disseminated beyond the Health Sciences Schools to many of the schools and colleges on campus. This has resulted in pilot awards for junior faculty in the Graduate School, the School of Social Work, and the College of Letters and Sciences, among others, as well as creation of innovative research partnerships.

Novel Multistep Review Processes for T1 and T2 Applications

Each T1 research application undergoes a technical review by Pilot Award Program staff and also a scientific peer review with scoring by a committee of up to three experienced reviewers. In general, the top third of the applications in the review cycle are forwarded to a multidisciplinary review committee for final prioritization based on scores, the collaborative teams formed, the focus on targeted research areas, and the plausibility for success, and generation of data optimal for a subsequent application for NIH or other funding.

Each T2 application also undergoes a scientific peer review by a committee of up to three experts. The review is then evaluated by a 'study section' comprised of members from the ICTR and from the community. Those applications with significant merit are then further evaluated by the External Community Review Committee which was created to include a community health-stakeholder voice in establishing the research agenda and making final research recommendations. Final funding recommendations are based, in part, on such criteria as significance of the research question to community health, priority of the research for the funding institutions, and evidence of community input into the application. The ICTR Executive Director and Senior Associate Executive Director review and approve all final funding decisions for both T1 and T2 applications.

Mentoring Awardee Progress and Tracking Award Outcomes

Each pilot grantee submits quarterly reports that allow program staff to monitor progress and offer assistance to manage unanticipated problems (e.g. recruitment challenges, need for additional ICTR resources, marshaling of additional scientists/mentors to the project). In addition, program staff also maintain contact with pilot award recipients well after their projects are completed in order to remain informed about their publications, extramural grants, contracts, or other accomplishments related to their pilot funding. This oversight provides 'rescue' of projects midstream, assessment of the common metrics for success on a continual basis, and insight to long-term success.

Table 1. Summary of program enhancements

Strategic process	Date	Results/intended outcome
Introductory 'brown bag' sessions	2009, round 3	PIs must identify and contact collaborators earlier in the process; applicants have a better understanding of grant mechanism
Mandatory letter of intent	2009, round 3	staff better prepared to assign reviewers, plan for proposals
Mandatory T2 'community abstracts'	2008, round 2	enhanced ability for community evaluation of projects; PIs to consider community impact/perspectives of research
Mandatory T2 grant writing workshops	2009, round 3	improved scores on proposals
Addition of qualitative/ mixed methods resource (T2 programs)	2011, round 5	3 pilots consulted with our qualitative/mixed methodologist in crafting their applications; one engaged in full collaboration – we expect this number to increase over time

Strategic Process Improvements

The monitoring of the Pilot Awards Program has resulted in many improvements in the process of application preparation and submission (table 1). To ensure the success of junior investigators in conducting their research projects, particularly those involving T2 research, ICTR has developed resources/programs such as: (1) educational (hands-on) workshops that focus on: (a) an introduction to qualitative data analysis; (b) the art and science of interviewing groups with a focus on group fundamentals; (c) applying theory in health behavior research; and (d) patient education interventions, most notably theoretical and operational considerations in designing a successful trial; (2) identification of faculty with expertise in qualitative and mixed methods analysis, who can support junior faculty in preparation of their applications and/or perform analysis of the funded research; and (3) 20 h of 'free' biostatistical support.

Award Types

The ICTR has provided in excess of USD 1.2 million annually to support pilot award funding. The type of pilot awards available has increased to align more appropriately with the educational/research needs of the junior faculty. The five different types of pilot awards currently available include:

1 T1 Pilot Awards: this USD 50,000 award for 12 months supports the spectrum of T1 from basic research to clinical trials and clinical studies, which includes

patient-oriented research that embraces innovations in technology and biomedical devices.

2 T2 Pilot Awards: this USD 50,000 award for 12 months supports research that translates knowledge into improvements in clinical practice and community health, as well as development of health care interventions that require a change(s) in individual, organizational, or system behavior.
3 T2 Community Collaboration Grants: this USD 200,000 award for 24 months is provided to junior or senior investigators who have a robust community partner/collaboration; the award requires a high level of community collaboration and inclusion of senior investigators with long-standing community relationships, both of which create more active participation of community research partners.
4 Cross CTSA Collaborative T2 Community Grants: the ICTR has provided pilot awards to investigators who have developed collaborations with researchers at the University of Minnesota, a recently-funded CTSA academic health center within our region; the grants, which focus on issues of importance to the communities of the region and include community participation, are cofunded by the participating CTSA institutions and range from USD 100,000 to USD 200,000 for 1–2 years.
5 Targeted Research Awards: many of the pilot projects have targeted specific research areas and are cofunded by a variety of partners (table 2).

Accomplishments

Over the initial 4 years of the program, the ICTR has supported 102 pilot awards (546 applications) in five rounds for a total investment of USD 5.41 million. Fifty-nine T1 and forty-three T2 pilot projects were funded at total costs of USD 2.78 million and USD 2.71 million, respectively. The majority were awarded to PIs who held the rank of Assistant Professor or Assistant Scientist; the funding has provided the opportunity for them to generate preliminary data for subsequent grant applications and publications. Many of the awardees were part of new collaborative associations between junior and senior investigators or with new interdisciplinary teams, which often included investigators at the University of Wisconsin-Madison and Marshfield Clinic; these collaborations represent an important shift toward integrating more diverse perspectives in the design and conduct of health care research. Funded PIs now represent nearly all of the academic units that partnered to form the ICTR, as well as academic units that were not initially involved.

The Pilot Awards Program portfolio mirrors the breadth of health interests at the University of Wisconsin-Madison and Marshfield Clinic, as well as throughout Wisconsin. The projects broadly encompass basic biomedical and comparative biosciences; human clinical interventions; and community-based effectiveness, intervention, and policy research. With the purposeful crafting of the requests for applications, ICTR has also supported research projects relevant to special populations, namely children and those who experience health disparities. The ICTR has leveraged local

Table 2. Contribution of cofunders to pilot grant awards

Cofunding entity	Year 3 USD	Year 4 USD	Year 5 USD
University of Wisconsin Comprehensive Cancer Center	47,900	49,977	50,000
Wisconsin National Primate Research Center	25,000	25,000	25,000
Coulter Foundation/Biomedical Engineering	24,955		
University of Wisconsin Waisman Center	25,000	25,000	24,972
Stem Cell Regenerative Med Center		50,000	98,882
Marshfield Clinic Research Foundation		50,000	25,000
Community Academic Aging Research Network			25,000
University of Minnesota		75,000	
Totals	122,855	274,977	248,854

resources to assist several T2 grantees in accessing a statewide community-engaged research infrastructure that includes research 'ambassadors' in African-American and Latino communities in Milwaukee, Wisconsin, and in multiple American Indian tribal communities throughout the state.

The return on investment of the ICTR Pilot Awards Program has been realized in extramural funding and peer-reviewed publications. The 35 pilot awards funded in in the first year of the CTSA were the basis for extramural grant funding in excess of USD 20 million, with a rate of return near 13-fold (pilot award funding: USD 1,567,274; extramural grant funding USD 20,108,405; 12.8-fold). Out of 41 total grant applications originating from these pilot awards, 18 were funded; of these, 10 were NIH awards.

In parallel, the research supported by the pilot awards funded in year 1 led to 32 peer-reviewed publications, resulting from 40 submissions. The full impact of grants and publications from years 2–4 are not yet realized, as completion of the research and analysis of the data have only been relatively recently completed or are in process.

The Pilot Awards Program at the University of Wisconsin has been remarkably successful, serving as a 'launching pad' for research careers of junior faculty, creating long-lasting research partnerships, and expanding the research portfolio at the University of Wisconsin, particularly the T2 activities of the faculty. As the second funding cycle for the University of Wisconsin CTSA grant begins, the ICTR leadership envisions robust continuation and further development of the Pilot Awards Program.

Acknowledgement

Preparation of this chapter was supported in part by grant No. UL TR000006 (A.S.) and UL1TR000427 (M.K.D.) from the NIH, National Center for Advancing Translational Sciences, Clinical and Translational Sciences Award.

References

1 Butler D: Translational research: crossing the valley of death. Nature 2008;453:840–842.
2 Woolf SH: The meaning of translational research: why it matters. JAMA 2008;299:211–213.
3 Zerhouni EA: Translational and clinical science – time for a new vision. N Engl J Med 2005;353: 1621–1623.
4 Zinner DE, Campbell EG: Life-science research within US academic medical centers. JAMA 2009; 302:969–976.
5 Rosenblum D, Alving B: The role of the clinical and translational science awards program in improving the quality and efficiency of clinical research. Chest 2011;140:764–767.
6 Califf RM, Berglund L, Principal Investigators of National Institutes of Health Clinical and Translational Science Awards: Linking scientific discovery and better health for the nation: the first three years of the NIH's Clinical and Translational Science Awards. Acad Med 2010;85:457–462.
7 Trochim W, Kane C, Graham MJ, Pincus HA: Evaluating translational research. Clin Transl Sci 2011;4: 153–162.

Scott C. Denne, MD
Indiana Clinical and Translational Sciences Institute
Indiana University School of Medicine, 410 W. 10th Street
Indianapolis, IN 46202 (USA)
E-Mail sdenne@iupui.edu

Alving B, Dai K, Chan SHH (eds): Translational Medicine – What, Why and How: An International Perspective.
Transl Res Biomed. Basel, Karger, 2013, vol 3, pp 74–81 (DOI: 10.1159/000343022)

Building Expertise in Translational Processes through Partnerships with Schools of Business

Steven K. Cheng · David M. Dilts

Knight Cancer Institute, Center for Management Research in Healthcare and Division of Management,
Oregon Health and Science University (OHSU), Portland, Oreg., USA

Abstract

Translational research in medicine is facing burdens stemming from an increase in the complexity of science, increase in partnerships across national and international collaborations, and reduction in the finite resources to support all research endeavors. Schools of business offer unique perspectives on translational processes because they address global challenges through research and teaching to transform ideas into successful practice-changing innovations. While there are multiple approaches to investigating translational processes using business management tools, this chapter will focus on three representative lenses: (1) process flows for mass customization, (2) knowledge supply chain, and (3) strategic management. Each lens yields the potential to significantly streamline the translational processes in healthcare for efficiency and effectiveness.

Medical advancements across all disease domains, coupled with the dissemination of inventions, are yielding great strides in healthcare, yet the cost worldwide is increasing at a pace that is higher than any country can sustain. In some counties, the required investments in biomedical research are increasing at a time when public funding is becoming static or even decreasing. With new treatments costing hundreds of millions of dollars to create, requiring decades to develop and bring to patients, and often having low success rates, leaders in biomedical research and healthcare are primed to examine other disciplines to find solutions to optimize efficiency and effectiveness. Compared to other industries, biomedical research and medicine are decades behind in the utilization of business infrastructure and operational management concepts. As stated in the April 2010 Institute of Medicine report, *A National Cancer Clinical Trials System for the 21st Century* [1], 'A new and novel approach is required to solve these well-known intractable problems, with application of the best minds in multiple disciplines (e.g. engineering, social science, management, and marketing).' Hence, the question at hand is: 'What can translational research learn from the expertise found in business schools?'

In this chapter we introduce three lenses from which to view the potential application of business management techniques to translational medicine: (1) process flows for mass customization, (2) knowledge supply chain, and (3) strategic management. Mass customization flows address removal of barriers that can enable translational research enterprise to be more effective while simultaneously allowing for high variability in output. Knowledge supply chain focuses on collaborative and cooperative opportunities that can facilitate the transfer of new knowledge, mechanisms, and techniques among 'suppliers', i.e. laboratories, clinics, and health systems. Strategic management concepts can bring discipline in establishing a well-defined approach for fostering key translational initiatives. The views through these lenses illustrate how business management techniques provide a critical foundation for building expertise in conducting translational science.

Three Management Lenses for Translational Research

Process Flows for Mass Customization

Process flows for mass customization utilize standard modules or subcomponents for traditional manufacturing and service operations while accommodating for unique specifications required by the 'customer.' This blend of customization and standardization recognizes and addresses infrastructural and administrative barriers that hinder performance, while providing success measures such as quality, speed of product delivery, customization, and new product introduction. Standardizing process flows does not mean building bureaucratic processes that stifle innovation. Rather, industries have discovered that the right level of standardization enables the generation of a greater variety of products with shorter product development cycles with the same amount of resources. For example, computer firms, like Dell Computers, are faced with the manufacturing requirements of creating 'personalized' computers that incorporate unique, customer-specified products without the degradation of time to the customer and at competitive costs [2].

Looking at translational research through the lens of process flow can uncover inefficient and wasteful steps, which can be eliminated with no impact on the safety or efficacy of the research. We have brought business school methods to one part of the translational research process – the oncology clinical trial. In this case, each trial was considered as a 'product' that requires processing at various 'jobs' that required customization prior to completion. In our investigation of government-sponsored late-phase multi-institutional oncology trials, nearly 500 unique process steps across more than 800 days were identified to transform an investigator's idea into a clinical trial ready for its first patient [3]. In-depth analysis of the clinical trial development

Table 1. Types of administrative barriers (adapted from [6])

Administrative barriers	Definition	Representative statements
Procedural barriers	policies, either formal or informal, that arise from the processes that may inhibit problem solving actions	'that's not my job'
Structural barriers	created when different participants in the process follow a different ordering of steps, which can lead to miscommunication and confusion	'that is not the way we do things here'
Infrastructural barriers	concern how the underlying system is designed and how it supports the interconnection of various aspects of the systems	'I would love to do that, but the system does not allow it'
Synchronicity	occurs when dependent processes are delayed because of the lack of coordination and proper scheduling	'I cannot do what I need to do until they get done doing what they need to do'

process revealed more than 30 cycles ('loops') and over 36 different decision-making points [3–5]. The culmination of this investigation across the entire clinical trial process including scientific, regulatory, ethics, financial, contracts negotiation, and local site activation was a series of process maps that spanned over 40.23m (1,584 inches) long by 1m (40 inches) high – more than the wing span of a Boeing 737–900 [4].

This research identified 'invisible' barriers that were uncovered from the process-flow lens. These barriers, while only 'invisible' in the sense that they were not studied, were classified into four categories: procedural, structural, infrastructural, and synchronicity [6] (table 1). Procedural barriers are policies, either formal or informal, that arise from the processes required to activate a study and that may inhibit problem-solving actions. Structural barriers are created when different participants in the process follow a different ordering of steps, which can lead to miscommunication and confusion. Infrastructural barriers concern how the underlying system is designed and how it supports the interconnection of various aspects of the systems. Synchronicity barriers occur when dependent processes are delayed because of the lack of coordination and proper scheduling.

These barriers can be resolved through various tactics. The act of showing the entire process to the involved parties raises the level of awareness of the intricacies of the system and stimulates almost immediate suggestions for improvement. Unnecessary duplication can be quickly identified by simply looking at the flows on a process map. Understanding both processes and timing of individual jobs/steps can uncover opportunities where barriers of synchronicity exist. At times, rearrangement of steps can ensure that dependent steps are done in the proper order and that critical paths for completion designed. Processes that can be done (or not done) simultaneously can be identified. Tracing the routing of a trial through the process highlights decision points. Each decision point determines subsequent routes that the clinical trial must follow and can induce the requirement for rework and repetition of pre-existing process steps. By correctly making early decisions, looping (i.e. repeating steps) and rework can be reduced.

This view also reveals possibilities to investigate how to schedule and 'load balance' the process flow. For example, once a promising research project is identified, resources can be aligned in anticipation of its future arrival at later steps in the process. In this way, the research is not delayed awaiting resources availability.

Knowledge Supply Chain

While mass customization focuses internally within an organization, supply chain research focuses externally to the interfaces between collaborators and partners. A supply chain is a collection of organizations that maintain local autonomy and decision-making capability, yet act jointly to fulfill customer requirements [7]. The management of such a supply chain involves coordination of material and information flows among various organizations involved in producing a specific good or service. In contrast, a knowledge supply chain emphasizes the integration of intellectual and knowledge capital among collaborators organized around a common objective. In today's knowledge society, knowledge is the key driver to innovation and productivity [8]. A translational research knowledge supply chain is composed of a complex network of researchers, academia, industry, regulators, health care providers, funders, and patients. Each element 'supplies' unique intellectual assets such as basic research knowledge, patents, formulations, and capabilities – all which much must be created, traded, shared, and integrated into the translational process.

There are multiple pressures for managing such knowledge supply chains. First, developing a new treatment is extremely costly, reaching over USD 800 million per drug [9]. Second, as researchers explore the genomic or proteomic basis of disease, no single institution or region of the world will be able to identify and accrue all the biospecimens and patients required for such studies. Third, as the pressure for globalization increases, local translational science policies and practices are more likely to be shared and standardized across borders.

Multi-institutional and international partnerships on key scientific discoveries are critical to capitalize on the growing potential for economies-of-scope [10] as compared to economies-of-scale. Traditional economies-of-scale occurs when the same drug is mass-produced for the general population and thus cost advantages are derived from the volume produced and sold. In economies-of-scope, efficiencies are observed when collaborators focus on a few key targeted areas and utilize collective expertise to gain knowledge and financial economies.

Translational research has a unique knowledge supply chain because knowledge and power are distributed across multiple participants, making coordination and integration more problematic. The participants in this system are decentralized since they are not bound to the same formal processes, and they may be under different loci of control. The players in the knowledge and value chain in medical research encompass many disciplines across different types of institutions. For example, with more than 27 different types of groups or participants involved in developing a government-sponsored phase III oncology clinical trial in the United States, the number of different combinations of types of groups caused high variances in the processes and sequencing in the supply chain. Clinical oncology research within the USA is reviewed across various organizations including (but not limited to): academic medical centers, industrial sponsors, patient advocate organizations, National Cancer Institute at the National Institutes of Health, the Food and Drug Administration, and the oncology community. All groups collectively have the similar goal of bringing the clinical trial to patients, yet each group individually may have different sets of mandates, objectives, and requirements when developing the trial. Additionally, each organization is composed of semiautonomous groups that are directed by different decision-makers within the system. Often an academic cancer center is administratively responsible to a hospital or medical system; thus, the manager of the clinical trial supply chain must interface with various groups or departments, such as the scientific review committee, institutional review board, office of technology licensing, budgeting, and contracts and grants management. While none of these participants is solely dependent upon one another for survival, the rapid development and successful completion of a clinical trial is crucial to the mission of the total institution; hence, they should work together for overall success. Organizationally, however, such fragmentation and decentralization of decision-making, coupled with mixed performance goals, results in the creation of a complex system that behaves in a nonlinear, dynamic, complex, and sometimes unpredictable manner.

Best practices from standardization across the knowledge supply chain could be incorporated to prevent procedural and other administrative barriers from occurring. For example, blanket agreements between industry sponsors and academic institutions, if initiated and implemented properly, would reduce the administrative expense of processing multiple agreements while streamlining transactions between partners through the reduction of nonessential steps. As global translational research becomes more prominent, standards across multiple international regulatory and oversight bodies could be made. This would yield organizational efficiencies, development of standardization to facilitate collaboration, creation of clear and transparent rules to reduce ambiguity, and the creation of a common multinational platform to aid in the translational research infrastructure, while still allowing for a degree of local autonomy and customization to local needs. Such local autonomy coupled with joint action is needed in translational research, where basic scientists, clinical researchers, pharmaceutical firms, government agencies, and communities must work together to quickly bring new approaches for prevention, diagnosis, and treatment of disease and improving health.

Strategic Management

Strategic management is a formal approach by which activities, courses of actions, and patterns of decision-making are linked to maximize capabilities for achieving a predetermined purpose and focus, that is, it focuses on effectiveness of the process rather than efficiency [11–13]. Strategic management emphasizes the development of macroscopic policies to ensure that research efforts will advance science and potentially produce therapeutic applications [14]. For example, government agencies must decide how best to provide funds and other resources to effectively implement health policies across multiple competing and complementary domains. Research institutions must select and exploit specific opportunities that carefully balance their expertise, foster collaborative research teams, and engage the academic community. They also need to decide how to reward individual and interdisciplinary achievements and efforts.

There are multiple models of viewing the strategic flow of research. Classically, management of research is depicted as either a linear flow from basic to applied research or as an ever-ascending staircase – each step representing a novel discovery or development building off another. Conversely, managing curiosity is critical for capitalizing on unexpected discovery. Bell Laboratories was able to successfully blend the various models for strategic flow of research where innovations ranging from the telephone exchange, transistor, and radio astronomy were discovered and seven Nobel Prizes were awarded to its researchers [15]. Strategic management ensures that when findings are discovered from preclinical and basic research, the organization and systems are ready to support progress through the translational pipeline.

At any given time, enterprises doing translational research are balancing a multitude of research opportunities, all of which are at various stages in development. Some basic research has a direct application and motivation for its effort. Pasteur's collaborator Emilie Roux wrote that the pursuit to find the rabies vaccine was motivated not merely because the disease was 'the most subtle and the most mysterious', but also it was 'the most frightening and dreaded malady' in the mind of the general public. Conversely, the creation of fiber optics and lasers were developed with no direct application in mind. It was not until Bell Laboratories linked and applied these technologies, along with semiconductors, that the amount of data transferred electronically was able to vastly increase [15]. This also led to the radical transition in networks (from analog to digital) that is the basis of all of today's digital communications.

One popular technique in strategic management is the use of portfolio management, which allows for a mix of different research types, characteristics, and potential impacts. In one product-development matrix approach, a portfolio can be divided into breakthrough projects, platform projects, enhancement/derivative projects, and sustaining projects (fig. 1) [14]. In the realm of oncology clinical trials, for example, a breakthrough trial (project) would focus on a totally novel drug, involving a novel pathway, which could dramatically change how people with a particular type of cancer are treated. A platform trial (project) could be one in which a drug that has been

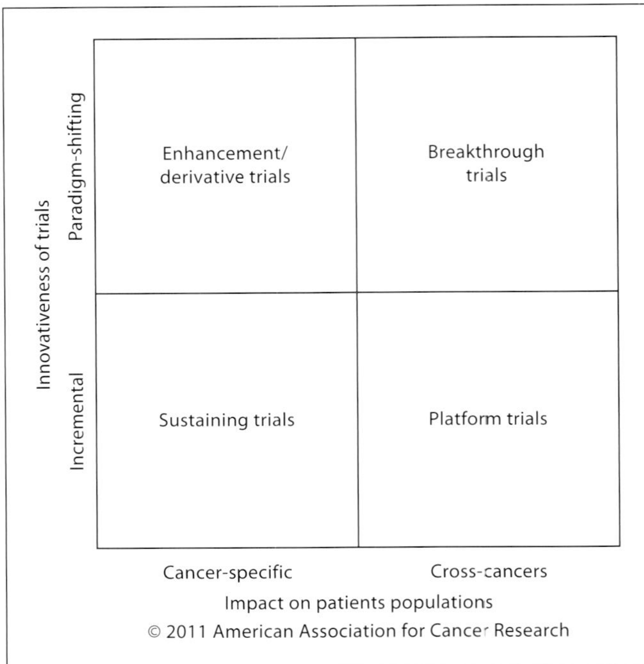

Fig. 1. Clinical trial portfolio matrix [14].

successfully used for one type of cancer is evaluated for use in a different type of cancer. An enhancement/derivative trial (project) could be a phase II trial that includes additional targeted biomarker screening to tailor cancer therapies and investigate outcomes. Finally, a sustaining trial would be one that focuses on fine-tuning dosages or treatment cycles.

Conclusion

The lessons that can be applied from other industries to translational research are not new. Calvin Tomkins [16], in 1963, described translational research conducted at Bell Labs as 'bringing the [country's keenest thinkers] together to bridge the gap between the best science of the academy and the important applications that a modern society needed.' Translational research has become a global activity requiring intricate collaboration and coordination across multiple environments and conditions. Such activity, coupled with the creation of a fast-paced 'discovery engine', comes with associated hurdles and barriers. However, many of these hurdles have been overcome in other industries and research settings by applying business concepts and theories. This chapter only highlights three management perspectives, but there are multiple additional lenses that can be applied to translational research where schools of business have expertise (e.g., project management, diffusion of innovation, and new product development). Translational researchers and institutional leaders are beginning to tap into this expertise to form productive interactions with schools of business. These tools and methods, which are only beginning to be utilized in biomedical and translational research, will soon be recognized as essential elements for driving efficiency and effectiveness in outcomes from the laboratory to the community.

References

1 Institute of Medicine. A National Cancer Clinical Trials System for the 21st Century: Reinvigorating the NCI Cooperative Group Program. Washington, Institute of Medicine of the National Academies, 2010.

2 Dell M, Fredman C: Direct from Dell: Strategies that Revolutionized an Industry. New York, HarperBusiness, 2006.

3 Dilts DM, Sandler AB, Baker M, Cheng SK, George SL, Karas KS, et al: Processes to activate phase III clinical trials in a cooperative oncology group: the case of cancer and leukemia group B. J Clin Oncol 2006;24:4553.

4 Dilts DM, Cheng SK, Crites JS, Sandler AB, Doroshow JH: Phase III clinical trial development: a process of chutes and ladders. Clin Cancer Res 2010; 16:5381.

5 Dilts DM, Sandler AB, Cheng SK, Crites JS, Ferranti LB, Wu AY, et al: Steps and time to process clinical trials at the Cancer Therapy Evaluation Program. J Clin Oncol 2009;27:1761–1766.

6 Dilts DM, Sandler AB: The invisible barriers to clinical trials: the impact of structural, infrastructural, and procedural barriers to opening oncology clinical trials. J Clin Oncol 2006;24:4545–4552.

7 Simchi-Levi D, Kaminsky P, Simchi-Levi E: Designing and Managing the Supply Chain: Concepts, Strategies, and Case Studies. New York, Irwin/McGraw-Hill, 2003.

8 Drucker P: From capitalism to knowledge society; in Neef D (ed): The Knowledge Economy. Woburn, Butterworth-Heinemann, 1998, pp 15–34.

9 DiMasi JA, Hansen RW, Grabowski HG: The price of innovation: new estimates of drug development costs. J Health Econ 2003;22:151–185.

10 Panzar JC, Willig RD: Economies of scope. Am Econ Rev 1981;71:268–272.

11 Mintzberg H, Ahlstrand B, Lampel J: Strategy Safari: A Guided Tour through the Wilds of Strategic Management. New York, Simon & Schuster, 2005.

12 Hax AC: Redefining the concept of strategy and the strategy formation process. Strategy Leadership 1990;18:34–39.

13 Porter ME: Competitive Advantage. New York, Free Press, 1998.

14 Dilts DM, Cheng SK: The importance of doing trials right while doing the right trials. Clin Cancer Res 2012;18:3–5.

15 Gertner J: The Idea Factory: Bell labs and the great age of American innovation. New York, Penguin Press, 2012.

16 Tomkins C: Voomera Has It! The New Yorker, September 21, 1963:49.

Steven K. Cheng, PhD
Senior Research Associate, Center for Management Research in Healthcare, Knight Cancer Institute
Adjunct Faculty, Division of Management, Oregon Health & Science University (OHSU)
3303 SW Bond Avenue, Portland, OR 97239 (USA)
E-Mail chengs@ohsu.edu

Alving B, Dai K, Chan SHH (eds): Translational Medicine – What, Why and How: An International Perspective.
Transl Res Biomed. Basel, Karger, 2013, vol 3, pp 82–88 (DOI: 10.1159/000343012)

Creating Ethical, High-Value Industry-Academic Partnerships

Scott Steele[a] · Kenneth J. Holroyd[b] · Terry J. Fadem[c]

[a]Office of Research Alliances and Clinical and Translational Science Institute, University of Rochester, Rochester, N.Y., [b]Center for Technology Transfer and Commercialization, Vanderbilt University, Nashville, Tenn., [c]Office of Corporate Research, Perelman School of Medicine, University of Pennsylvania, Philadelphia, Pa., USA

Abstract

The number of industry-academic partnered relationships has grown over time to encompass most companies and virtually all of academia. This should result in advancing new ideas more rapidly to patient care. These relationships also potentially generate significant benefits to both parties enough so that care must be taken to ensure that the credibility of each remains intact. This requires diligence in establishing collaborations that includes active management of them on the part of the university, while industry must equally manage the benefits they offer to individuals and institutions in a transparent manner. In the end, the responsible objective must be informed decision-making by providers and their patients or care-givers for the therapies they choose.

Industry and academia have historically existed in a codependent relationship. Universities strive to generate the knowledge necessary to enable medical problems to be identified and solved, while commercial entities are economically incented to utilize this knowledge to produce and deliver solutions.

For academia, this means that in order to deliver the university discovery of insulin, for example, to people suffering from diabetes, it was necessary for a company to produce and distribute the product so that it could be available to patients worldwide. On the other hand, for industrial innovations in medicine – from drugs to devices – academic partnerships are equally necessary to determine which patients benefit, why they benefit, and how to manage the health of the treated patient.

Throughout much of the twentieth century, companies were located in close proximity to universities so that they could directly benefit from collaboration with academic scientists and clinicians [1]. These relationships were relatively transparent, as were the benefits. As the local partnerships have evolved into international collaborations driven by strategic interests and access to specialized science, relationships have become more opaque over time. This reduced visibility has resulted in an inability to take the lessons learned from one partnered relationship to another or from one company or even one university to the next, hampering our collective ability to continue to improve collaborative relationships. Our aim in this brief essay is to provide insight into the process of creating effective and ethical industry-academic relationships.

The Current Situation

Research and development are often viewed as occurring in several stages, from basic or discovery research, through clinical and translational research, to implementation and community engagement. Academic-industry partnerships can be found at each of these stages and they vary in scale and complexity. The University Industry Demonstration Partnership, convened by the US National Academies, recently released a report that discusses a continuum of public-private partnerships [2]. These partnerships are viewed as taking many forms, but several key categories of activity are included in most: student-oriented engagement, involvement with researchers, access to resources, involvement with centers of excellence or specific schools, and economic development initiatives. Additionally, within each activity category, partnerships can vary from tactical one-time interactions to more multidimensional strategic collaborations that may encompass basic science on mechanisms through to outcomes studies [3].

Across the research and development spectrum, US federal investments are the critical source for support of basic research performed at academic institutions, accounting for approximately 60% of academic spending on science and engineering research and development in fiscal year 2009 [4]. At the same time, the National Institutes of Health (NIH) is the most significant funder of all the federal agencies, accounting for approximately 65% of the total federal academic research and development contributions [4]. Industry's percentage of funding for academic research and development declined steeply after the 1990s, from more than 7% in 1999 down to about 5% in 2004, but has seen a 5-year increase to about 6% in 2009 [4]. In particular, industry engagement with academia is essential for the later stages of the research and development pathway. Key opinion leaders are often resident at major academic centers, and they are recruited by industry for their expertise as well as their potential influence once a product is ready to market. The academic institutions themselves offer further credibility by lending their brand name to clinical studies by way of sponsored research agreements.

Currently, about 80% of industry-sponsored research collaborations support evaluation in human subjects and the remaining 20% support preclinical evaluation and product discovery. This investment by industry in a single academic institution could be anywhere from USD 10 million to greater than USD 100 million paid to the institution in any given year. In addition, a small but increasingly important industry investment in academia supports precompetitive projects of interest to multiple companies in an industry. This has potential for future expansion, in part as the US government incentivizes universities to participate and develop such programs in cooperation with industry. Examples of precompetitive projects include biomarker development, target validation, and regulatory science.

The amount of industry sponsorship of university partnerships follows the research and development budgets of the life sciences industry, with the largest number of projects, and most monetary support, provided by the pharmaceutical and biopharmaceutical companies, followed by the medical device, and then diagnostics companies. Partnerships emphasizing research and development of health care information technology, clinical decision support, and improvements in health care services and related health outcomes are relatively rare.

The vast majority of financial support for sponsored research comes about as a direct contact between company translational scientists and their counterparts in universities. A small minority arises from interactions initiated by leadership of the companies and universities, although these strategic relationships are usually more significant in scope. Similarly, a sustained alliance management initiative across years and multiple projects between a university and a company is less common, but when present is quite often productive. Universities are increasingly employing full-time 'alliantists' to encourage and sustain such partnerships, and to match the important function of the industry alliance management function.

Colocation of industry near university scientists is a growing trend and has been helpful, particularly for partnerships initiated between a company and university that are in close proximity. Similarly, most high-value, sustained translational science partnerships have been created when the university and company are located in the same city or region, although there are exceptions.

One hundred percent of industry translational scientists have experience in academia since they obtained their professional degrees in universities, but we estimate that less than 1% of university-based translational scientists have spent any time in industry. This often results in unmet expectations by both the industry and academic scientists since they may not have effectively communicated their respective primary objectives to one another. For industry scientists, the bottom line is cost-effective development of products, such as new drugs or devices. For the academic scientist, publishing high-quality papers, along with advancing the research of his/her laboratory is paramount.

Risk sharing and investment by the universities specifically for a partnership project are not common since universities rarely have the resources to invest beyond needed infrastructure for research. This has started to change as companies have initiated risk/reward sharing models designed to attract innovative academic research with the promise of significant future gain. However, risk mitigation is something unfamiliar to academia, which values all knowledge, since positive as well as negative results generate papers. For academia, the potential return on dedicating too much effort to meet any corporate objective may remain just that – unrealized potential.

The Benefits of Academic-Industry Partnerships

The headline was simple: 'CHOP doctors find success with gene drug' (where CHOP is Children's Hospital of Philadelphia). The success was fostered by an industry-academic alliance between Pfizer and the University of Pennsylvania where a joint research collaboration committee debated and then agreed to fund a proposal made by CHOP to try the lung tumor drug crizotinib, in children suffering from a rare form of neuroblastoma [5].

Productive university-industry partnerships have the ultimate goal to help patients and broadly improve public health. Such collaborations can take many forms across the spectrum of the research and development pipeline. Strategic and tactical alliances benefit the goals and missions of each partner, but also serve a greater public good. Successful collaborations can address smaller technical demands, as well as grand challenges that could not be addressed without combining the resources, expertise, and knowledge of these partners. This has been true historically and remains a critical priority given the current financial climate and growing global competitiveness.

These partnerships benefit society at many levels, with some more apparent than others. University-industry partnerships continue to facilitate the essential discovery of knowledge, and ultimately they impact the dissemination and application of that knowledge to improve public health by developing new and improved solutions to prevent, detect, diagnose, treat, and cure diseases. In the process, academic institutions fulfill their mission of education/training and creating and disseminating knowledge, and companies are able to meet their goals of providing innovate and competitive products, while creating increased value for their investors.

As Gary Pisano of Harvard Business School and others have noted, life science product development is distinct from many other areas of technology product development and investment because of the tremendous uncertainty that infuses each step and component of the discovery and development process. University-industry partnerships allow the following emphases to work to the mutual benefit of both parties. Cutting-edge scientific knowledge can help lower the risk and reset priorities for product development. Many product discovery and development questions have at their root scientific questions of intrinsic merit, for which proposed studies may attract peer-reviewed government funding.

Government funding of scientific questions which intersect product discovery and development can have a double payoff: decreasing the risk of industry investment in product development and adding to important scientific biological and chemical knowledge. Examples include studies of the chemical biology of 'undruggable' molecular targets, and the development of chemical and biological probes for imaging new molecular targets, which in each case can become lead compounds potentially resulting in new drugs. Lastly, the experiences of cross-disciplinary training in product discovery and development thinking and methods for academic translational scientists will allow greater incorporation of these activities into their other academic projects. Hopefully over time, this will lead to better academic translational science success. As noted, industry thinking and methods for roughly 99% of academics are largely a 'black box' of experiences at the current time.

University, industry, and government research consortia have been used in the past as one mechanism to enhance coordination and pool resources to address grand challenges to national security and competitiveness. The consortium formed in 1987 among 14 semiconductor companies and the federal government and known as Semiconductor Manufacturing and Technology (SEMATECH; http://www.sematech.org) is a primary example of a consortium focused on semiconductor research and development that began with crucial support from the federal government. This was initially established to address the critical lack of US competitiveness in the semiconductor industry and was viewed as a national security issue as well. Another promising model comes from the network of academic institutions established through the NIH's Clinical and Translational Science Awards (CTSA; http://www.ctsaweb.org/). The current NIH CTSA program provides funding to support a consortium of 60 biomedical institutions focused on a common goal to improve public health through enhancing the efficiency and effectiveness of clinical and translational research. A number of national CTSA committees have been established to enhance consortium-wide initiatives, including the establishment of a public-private partnership committee designed to develop programs and procedures to encourage private sector engagement with the CTSA consortium. Indeed, the NIH has highlighted this as an area of focus [6], and with the establishment of the new National Center for Advancing Translational Sciences (http://www.ncats.nih.gov/) the NIH has further enhanced programs focused on public-private partnerships (http://www.ncats.nih.gov/research/reengineering/rescue-repurpose/therapeutic-uses/therapeutic-uses.html).

For each individual and institution, ethics exist in a matrix of interests. Maintaining rigorous ethical standards in university-industry partnerships is critical for success. The most fundamental ethical challenge is the potential financial gain that can accrue to academics who are in a position of institutional, professional, and personal patient trust. However, there are many other potential influences that raise ethical questions, such as personal commitments, beliefs, politics, and family indicated in the figure 1.

Corporate interests are well understood – the focus is to continue to invest in growth of revenue and earnings [7]. University sensitivities are appropriately of most concern once product development moves to human subject evaluation, and further during the final phase of testing before regulatory marketing approval. Universities are directed to have a rebuttable presumption that human subject testing can be performed at other institutions (i.e. the universities need to justify why and how investigators with financial interests in a particular area of research can be involved with this research at their own institutions). These concerns have always been justifiably present. Similar institutional conflicts of interest arise during any technology commercialization process, whether with government or industry sponsorship, as universities may profit from such activities under the US Bayh-Dole Act and similar international legislation.

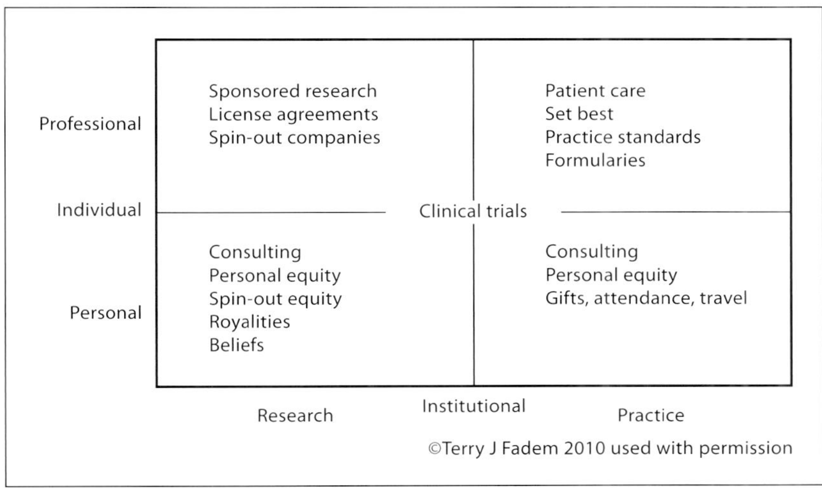

Fig. 1. Ethical challenges.

One challenge is to ensure institutional policies are clear and consistent. This is incumbent on both sides. The objective that should be adopted is to alleviate concern by avoiding any potentially hidden conflict. In an environment of growing regulatory burdens, all government agencies want to foster more public-private partnerships to promote innovation and the development of new vaccines, drugs, and therapeutics.

Conclusions and Recommendations

Industry partnerships with universities and government, administered through corporate-sponsored translational research, reaches tens of billions of dollars per year in the USA alone. While poorly documented in quantity, the majority of the partnerships involve clinical evaluation of new therapies, and a lesser amount discovery and preclinical evaluation. Pharmaceutical company partnerships currently outnumber partnerships with firms developing devices or diagnostics. Health care information technology and services are new areas of potentially fruitful partnerships that will grow in influence as the costs of maintaining the continued stream of new technology is confronted by fiscal realities.

Partnerships in precompetitive activities such as biomarker development, target validation, and regulatory science are other promising future areas of expansion. In addition, formal business management of partnered relationships has enabled many institutions to benefit effectively. These are, after all, business relationships that have as their benefit potential improvement to patient care along with cost containment. Throughout the partnership process, the highest ethical standards, and management of conflicts of interest, must be maintained.

Outcomes should be measured by expansion of the magnitude of the activity and demonstrable increasing success of moving new products to the marketplace. An annual international survey of translational corporate sponsored research and partnerships between industry and universities, analogous to the annual Association of University Technology Managers survey for technology commercialization, should take place to document the progress.

The authors' recommendations for establishing and maintaining high-value successful academic-industry partnerships include:

- Placing patients first in every collaborative relationship – the potential patient benefits of the partnership should be clearly identified up front.
- Transparency – openly disclose all relationships, their value to individuals and the institutions.
- Improve integration – increase the opportunities to comingle university and industry scientists in a way that provides each with an opportunity gain the perspective of the other.
- Effectiveness – collaborate rather than 'hire'; industry can generate far more effective relationships by approaching each university and the scientists with whom they work as collaborators rather than contract research providers.
- Expand the nature of collaborative research relationships across the full breadth of academic capabilities including health care information technology, clinical decision support, and improvements in health care services and related health outcomes management.
- For academia in particular – manage corporate collaborative relations as business partnerships that have attendant responsibilities.

References

1 Macgarvie M, Furman JL: Working Paper 11470, Early Academic Science and the Birth of Industrial Research Laboratories in the U.S. Pharmaceutical Industry. National Bureau of Economic Research, Cambridge, 2005. http://www.nber.org/papers/w11470.

2 Partnership Continuum: Understanding and Developing the Pathways for Beneficial University-Industry Engagement. University-Industry Demonstration Partnership. http://sites.nationalacademies.org/xpedio/groups/pgasite/documents/webpage/pga_069334.pdf. p. 7.

3 Hughes B: Pharma pursues novel models for academic collaboration. Nat Rev Drug Discov 2008;7: 631–632.

4 U.S. National Science Foundation's Science and Engineering Indicators 2012, Arlington, VA. http://www.nsf.gov/statistics/seind12/pdf/c05.pdf.

5 Flam F: CHOP cancer doctors find success with gene drug. Philadelphia Inquirer, May 18, 2012, posted on philly.com 10:22 am.

6 Portilla LM, Alving BA: Reaping the benefits of biomedical research: partnerships required. Sci Transl Med 2010;2:35cm17.

7 A marriage of convenience. Nat Med 2012;18: 469–470.

Scott Steele, PhD
Research Alliances, Public-Private Partnerships, Clinical and Translational Science Institute
University of Rochester, 265 Crittenden Blvd, Office 1.208, Box CU 420708
Rochester, NY 14642-0708 (USA)
E-Mail scott.steele@rochester.edu

Alving B, Dai K, Chan SHH (eds): Translational Medicine – What, Why and How: An International Perspective.
Transl Res Biomed. Basel, Karger, 2013, vol 3, pp 89–97 (DOI: 10.1159/000343014)

Increasing the Efficiency and Quality of Clinical Research with Innovative Services and Informatics Tools

Mary L. Disis[a] · Peter Tarczy-Hornoch[a] · Bonnie W. Ramsey[b]

[a]Institute of Translational Health Sciences, University of Washignton and [b]Seattle Children's Research Institute, Seattle, Wash., USA

Abstract

A national investment in clinical and translational research has resulted in an analysis of the gaps that prevent the rapid translation of discoveries into therapies. By identifying the challenges facing clinical and translational researchers, best practices for streamlining resources and the development of novel informatics tools will aim to close the gaps. This chapter will use a case study of the re-engineering of the Cystic Fibrosis Therapeutic Development Network as an example of the power of metrics and continuous process improvement to accelerate the completion of clinical trials in rare populations and improve study quality across dozens of clinical sites. Further, complex trials, especially those that utilize intensive correlative laboratory studies, require equally complex data management and analysis tools. The development of a national biomedical informatics network via the National Institutes of Health-funded Clinical and Translational Science Award program has stimulated the rapid generation of open-source tools and robust commercial products to aid in assessing data from electronic medical records or in directly managing large newly generated data sets. By publishing standards and developing open-source informatics products for clinical research, the translational research revolution initiated in the USA will spread worldwide.

Clinical and translational research requires tools and services that are often more complex and expensive than those needed to conduct basic research. Clinical research must be supported by highly trained staff proficient in both medical and regulatory issues. Numerous complex databases, holding clinical and laboratory data, must be integrated to understand the role of specific biologic processes in mediating or predicting the course of a disease. These complex tools and services are often institutionalized or shared across multiple groups so as to improve affordability. Some critical tools, especially in the field of biomedical informatics, are prohibitively expensive for the typical academic investigator.

It is imperative that institutions and networks provide a suite of resources and tools to accelerate the clinical and translational research process, as well as improve the quality and safety of clinical research. Although the investment in such systems is significant, we will discuss how resources can be established under the principles of continuous process improvement (CPI), thus increasing efficiency and lowering costs of translational research without impacting quality. Intensive study of biomedical informatics processes across multiple institutions in the Clinical and Translational Science Award (CTSA) program, funded by the National Institutes of Health (NIH), will transform our current use of informatics tools. The CTSA is focused on developing best practices and off-the-shelf products that may deliver comprehensive informatics solutions to clinical and translational investigators worldwide.

Case Study: Re-Engineering a Clinical Trials Network

Clinical trials networks (CTNs) have become critical for the advancement of medical science and the evolution of medical care. In many disease settings where populations are limited (i.e. many genetic diseases) or the outcome is a rare event, CTNs provide a sufficient population base and centralized resources to ensure successful study completion within a reasonable time period [1]. The critical aspects of successful networks have been a well-established coordinating center that maintains communication across all components of the network and monitors the studies to ensure timely enrollment and completion. In 1987, the MacArthur Foundation prepared a report outlining the advantages and disadvantages of studies conducted under collaborative groups or networks [2]; the findings in the report are still relevant 25 years later. CTNs provide a larger patient pool, may be able to identify site differences, enhance medical science at the sites, provide training opportunities to young investigators, and benefit from broad expertise across the CTN in decision-making. The disadvantages are potential costs, inefficiencies of multiple institutional requirements, potentially slow decision-making, and lack of scientific recognition of individual investigators.

A case study of a CTN, the Cystic Fibrosis Therapeutics Development Network (CF-TDN), illustrates an attempt to eliminate the previously stated disadvantages by developing strategies to decrease costs and increase efficiency as well as standardize quality across sites. The CF-TDN was started in 1997 with the goal of developing new therapeutic approaches for the treatment and potential cure of cystic fibrosis [1]. As the network has grown to 77 sites, the CF-TDN has emphasized efficiency and cost awareness by investing in centralized Web-based data systems to conduct trials and closely follow clinical study metrics data. Each CF-TDN research site receives an infrastructure grant from the Cystic Fibrosis Foundation, which supports the network, to help fund a research coordinator and a small portion of an investigator's salary. To receive these infrastructure funds, CF-TDN sites must enter their metrics data into the database on a quarterly basis. These metrics data include number of days from site receipt of final protocol to institutional approval, contract approval, first patient screened at the site, and most importantly, the number of patients enrolled per site per study. These tools and processes have led to a culture of CPI [3]. The use of these tools and processes will be further described below.

How Clinical Study Metrics Data Are Used within the Network

Cystic fibrosis clinical research teams receive their site's metrics data providing them the opportunity to identify key barriers that may allow them to partner with other teams within their organization to facilitate process improvement (i.e. institutional review board or contracts offices). Sites are encouraged to develop innovative approaches to impact metrics and track subsequent improvements, such as the use of CPI programs such as Toyota Lean or Six Sigma. The data are then used to evaluate sites every 2 years for continued participation in the network. Historically, those sites with performance data in the lowest tenth percentile do not remain in the network, but have the option to reapply at the time of the next 2-year review. Finally, the data are used to track on an annual basis the overall performance of the CF-TDN in the conduct of clinical research studies. The most recent data analysis indicates that between 2009 and 2011, the median time from final protocol to first subject screened across all studies within the network decreased from 209 days to 182 days – an improvement of almost 1 month.

How Benchmarking Is Used to Identify Key Success Factors

The metrics data have revealed significant variability in performance, both in start-up timing and in enrollment success, across the 77 sites. Data for study start-up is shown in figure 1. Consistent with the principles of quality improvement, the CF-TDN are using these data to reduce variability and promote best practices across sites. To benchmark best practices, nine CF-TDN sites located at academic institutions with CTSAs were identified based upon exceptional clinical trials metrics. These metrics were derived from 58 multicenter clinical trials conducted between 2007 and 2011. These nine sites were visited and information regarding best practices and key success variables was collected by interview using a standardized set of interview questions [4]. The key success variables that were identified across all nine sites fall into three discreet areas including: efficient processes, good people (hiring, training, motivating), and a culture of research at the institutional level, at the interface between clinical care and clinical research, and in engaging patients to participate in clinical studies (table 1). Other important contributing factors to success included open communication across the research team and with the institution, optimal informational technology to support work processes, and utilization of robust, timely, and transparent measurement to drive improvement.

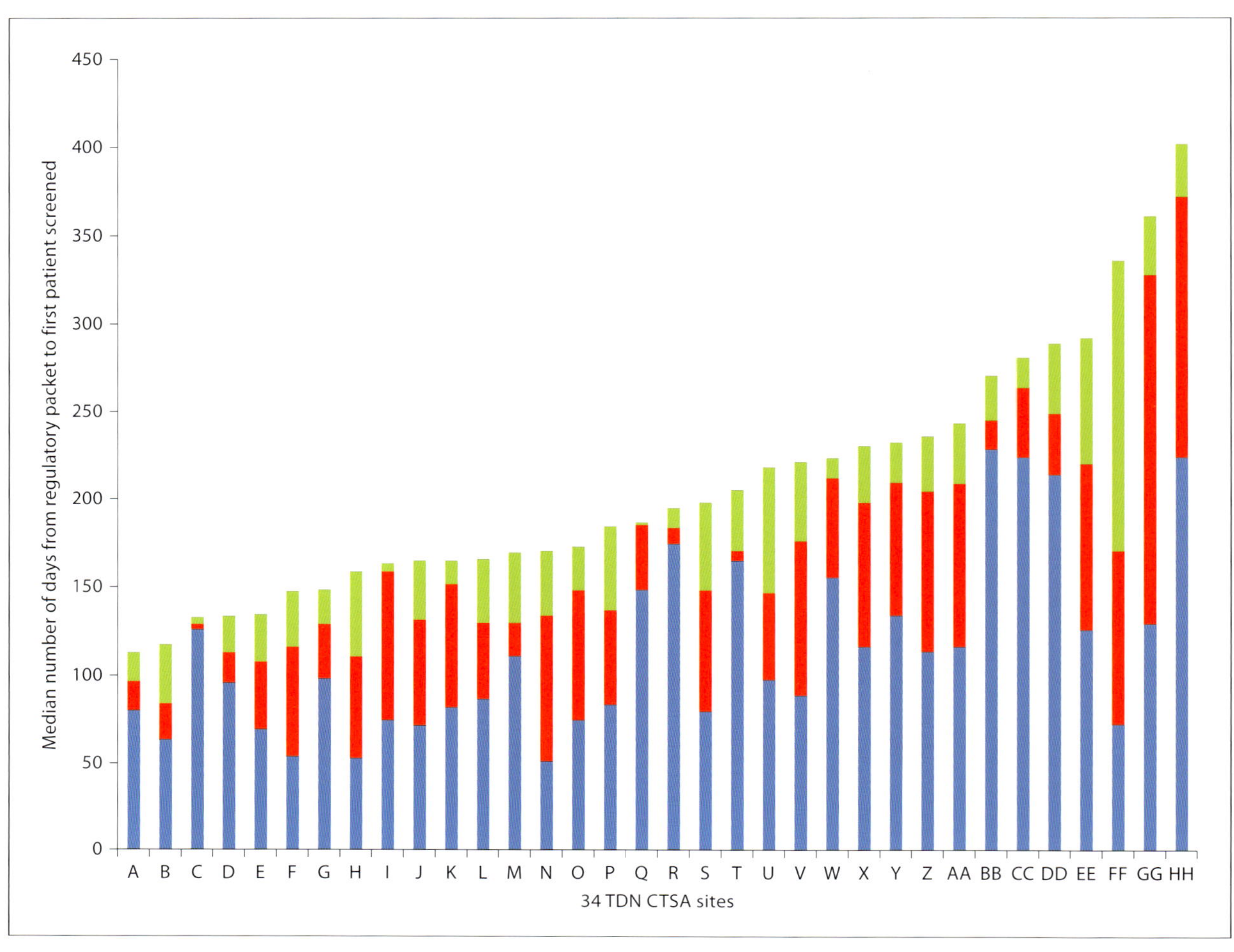

Fig. 1. Variability in the timing for study initiation and first patient enrollment across 34 institutions that are part of the CF-TDN. A composite of median start-up times across 34 institutions is displayed in the figure. The metrics were derived from 58 completed multicenter trials enrolling patients with cystic fibrosis over a 3.5-year period from 2007–2011. The x-axis is individual sites and the y-axis is the median time in days from the site's receipt of the regulatory packet (including the final protocol) to screening of the first patient for study enrollment. The total median time is divided into 3 components: median time from regulatory packet to IRB approval (blue), median time from IRB approval to site activation (red), and median time from site activation to first patient screened (green).

How Benchmarking Data Is Leveraged to Develop Tools and Training to Promote Quality Improvement throughout the Network

An additional objective for the qualitative research project was not only to conduct the benchmarking work and summarize the best practices discovered, but also to provide clinical research teams with training on conducting quality improvement activities and tools to assist them in establishing better practices. To that end, the

Table 1. Key elements of high-performing sites

Research team components	Institutional components
Multilevel leadership team structure	Mature culture of research at the institution
Collaboration for effective teamwork	Outstanding communication
Well-trained staff with longevity in position	Effective use of information technology
Customer service orientation (both team and institution)	Optimal sequencing of work processes
Thoughtful hiring practices to find the 'right' person	Strong institutional culture of CPI
	Use of transparent metrics to drive improvement
	Financial systems that support a business-like approach

CF-TDN has created a self-study curriculum entitled eQUIP-CR™ (electronic Quality Improvement Program for Clinical Research) as a Web-based system that is amenable to self-study. The system includes a detailed program assessment form (to identify areas for improvement) and instructions, templates, and clinical research specific case studies for process improvement techniques such as root cause analysis and process mapping [5], as well as a training plan for clinical research staff, tools for improving hiring [6], and sample subject satisfaction and research awareness questionnaires. The curriculum is supported by a webinar-based training and by a coaching program.

Information Technology Will Drive Both Innovation and Quality in Clinical and Translational Research

Acceleration of the translational research process will require rapid access to a variety of data. Moreover, CPI is dependent on a continuous flow of accurate data so that decision-making and process change can be seamless. The critical role of biomedical informatics in facilitating clinical and translational research was recognized by the NIH CTSA program that required biomedical informatics cores to be established at CTSA sites. Over the last several years, biomedical informatics researchers and the CTSA biomedical informatics cores have identified a series of recurring informatics needs of clinical and translational researchers. Initial qualitative research, adapting methods from clinical informatics, identified a number of broad themes (table 2) [7–9]. Some of these themes are human and organizational issues that impact on the success or failure of informatics interventions: communication, common workspace, and environmental factors (i.e. administrative and management structure, institutional support,

Table 2. Common translational informatics needs to ensure research acceleration

Environmental needs	Technology needs
Improved communication between researchers and developers	Managing, integrating, and analyzing large data sets
Common workspace to facilitate in-depth understanding of biologic and informatics intersections	Need of a comprehensive suite of tools across the spectrum of clinical and translational research
Institutional support for systems, technical personnel, and training	Interoperable systems to facilitate data exchange
Training of investigators to support self-sustainability	Data standards and security

technical support, training, funding). Some themes were particular to the biomedical informatics tools themselves, such as the challenges of using general purpose tools for data management (e.g. spreadsheets), lack of support for managing and analyzing large data sets, and need for the following: a suite of core services and tools, data integration and analysis, security structures, improved metadata and data provenance, interoperability and data exchange, and data standards. One can cluster the recurring biomedical informatics needs and tools developed across CTSAs around the different steps in the development and execution of a translational and clinical research project: hypothesis generation, study initiation, and the actual execution of the research project.

How Bioinformatics Is Critical in Accelerating Hypothesis Generation

With the growth in use of electronic medical records (EMRs), a wealth of clinical data is available in electronic form which can significantly aid in hypothesis generation. Unfortunately, EMRs are not designed to query or extract data for research purposes; thus, clinical and research data repositories have been built to enable investigators to extract medical data ranging from single institution solutions (e.g. the STRIDE system at Stanford [10], open-source solutions (e.g. the i2b2 open source repository [11]), and commercial solutions. Challenges with querying this complex data means researchers generally require assistance in extracting the data they need. Of note, the cost/effort is still substantially less than manual chart abstraction. Furthermore, since EMRs today contain large amounts of unstructured narrative text (e.g. admission history and physical), up to 50% of the data researchers need for a research question requires natural language processing or text mining to extract the information. As an example, in most health systems, smoking history is not a discretely coded data element, requiring a researcher to gather than information by reading the social history. Moreover, in order to rapidly acquire adequate sample sizes for studies, it would be optimal to integrate data across institutions. Such integration would require systems that 'talked' with each other and common language elements to describe existing data. The CTSA program is playing a key role in developing national and international standards and coordinating with other federal efforts led by the Office of the National Coordinator for Health Information Technology.

Another important use of clinical data repositories has been to aid researchers in accrual of subjects and/or biospecimens. If the eligibility criteria for a study can be expressed as a query against a data repository, then a tool could be developed that (subject to institutional review board oversight and approval) continually monitors clinical systems for potentially eligible subjects. Automatic flags built into the system could bring these individuals to the attention of a study research coordinator. Such custom tools have been implemented at specific CTSA institutions. If eligibility criteria depend on discrete computable data in the data repository (e.g. age, medications, diagnoses, lab values) then these systems work very well. If criteria depend on data that is often in narrative form (e.g. family history) or not captured (e.g. specific environmental exposures), then these systems work less well. In response to these challenges, investigators at Vanderbilt University created a tool called Research Match (www.researchmatch.org), which allows potential subjects to register on a website and volunteer for human studies. After local institutional review board approval of the proposed research projects, volunteers and researchers can be connected to determine if specific protocols are a fit for the volunteer based on his/her research profile. Research Match is a national effort and has enrolled thousands of volunteers while raising awareness about the importance of clinical trials.

In terms of biospecimen acquisition, the same general model of eligibility criteria expressed as queries identifying subjects pertains and a number of institutions have developed custom systems to support this type of workflow. For some studies it is useful to obtain discarded deidentified biospecimens (e.g. residual blood specimens that are a byproduct of usual care) with only nonidentifiable clinical data linked to samples. In this case, the option for EMR review to verify eligibility does not exist and the system works only if the eligibility criteria can be expressed as a query. Patients will, prospectively, need to be asked if they 'opt in' or 'opt out' to allow their samples to be used for research purposes. An example of such a study might be one which uses blood specimens from women between the ages of 18 and 40 with newly diagnosed breast cancer to develop assays for cancer biomarkers in serum. For prospective specimen collection linked to identifiable information or requiring capture of study specific specimens, patient consent is required. In this case, EMR review of eligibility prior to obtaining study relevant specimens or using discarded specimens is possible since the patient information is known.

How Bioinformatics Can Aid in Study Initiation and Design

In the exploratory phase of translational research, investigators look at the feasibility of conducting a study and try to find relevant expertise to assist in areas of study design, correlative science, and recruitment of patients. Biomedical informatics tools that have been developed to meet needs in the exploratory phase include tools to find potential collaborators, open-source and commercial deidentified clinical data repositories, and query tools that allow for cohort identification, power calculations, and estimation of subject and/or biospecimen accrual rates and times. For example, how many patients with a new diagnosis of multiple sclerosis are seen each year at a single institution? For rare diseases it becomes necessary to query these types of aggregate counts not just at one site, but across multiple sites. A pilot system is now in operation in the CTSA program [12] that includes 10 sites and 15 million patients for aggregate count querying by diagnosis. In addition, EMR data warehouses such as i2b2 (Harvard) and Amalga (Microsoft) allow data mining as described above.

How Bioinformatics Can Improve Data Quality in Ongoing Studies
During the execution of studies, researchers need to capture data specific to the study and not available from other electronic sources. Examples of this type of data include questionnaires administered to study subjects or disease/study-specific detailed medical or family history. For large-scale studies and for drug studies for regulatory approval clinical trial management systems (CTMS) have been developed and are available commercially. Since these larger studies typically are multi-institutional, CTMS are often centrally administered and remotely accessed by sites resulting in an efficient central capture and management and analysis of data. Unfortunately, in a centralized system, individual study sites inevitably copy data from electronic sources such as the EMR into the CTMS, resulting in duplicative work. Individual institutions have also purchased one of the multiple commercial CTMS systems and interfaced the software with local billing and EMR systems to minimize duplicate data entry. The high cost and complexity of these systems has limited their adoption. More commonly, investigators are using easier-to-adopt tools with more limited functionality focused on electronic data capture. A variety of tools, both commercial and open-source, exist for electronic data capture for small-scale clinical trials [13]. One very widely adopted basic tool is the open-source REDCap clinical data management system from Vanderbilt [14], which is both a workflow methodology and an adaptable tool for electronic data capture. The use of REDCap in the conduct of clinical trials has become a standard across CTSA institutions.

Conclusions

Clinical and translational research requires numerous resources and tools, covers a broad spectrum of disciplines, and requires analysis of complex data sets. Optimization of systems and development of novel informatics solutions requires multiple groups to collaborate, bringing diverse expertise and radical thinking to critical issues. The CTSA network provides a forum for national experts to tackle the fundamental problems that prevent acceleration of the clinical and translational research process. Solutions supported by the network are already being applied in research programs worldwide.

Acknowledgements

All authors are supported by NIH grant UL1TR000423.

References

1 Goss CH, et al: The cystic fibrosis therapeutics development network (CF TDN): a paradigm of a clinical trials network for genetic and orphan diseases. Adv Drug Deliv Rev 2002;54:1505–1528.
2 Kraemer HC, Pruyn JP: The evaluation of different approaches to randomized clinical trials. Report on the 1987 MacArthur Foundation Network I Methodology Workshop. Arch Gen Psychiatry 1990;47: 1163–1169.
3 Marshall BC, et al: Cystic fibrosis foundation: achieving the mission. Respir Care 2009;54:788–795, discussion 795.
4 Pawson R, et al: Realist review – a new method of systematic review designed for complex policy interventions. J Health Serv Res Policy 2005; 10(suppl 1):21–34.
5 Liker JK: The Toyota Way. New York, McGraw-Hill, 2004.
6 The Dartmouth Institute Microsystem Academy. Clinical Microsystems Toolkits: Workforce Development. 2012. http://clinicalmicrosystem.org/toolkits/workforcedevelopment.
7 Anderson NR, et al: Issues in biomedical research data management and analysis: needs and barriers. J Am Med Inform Assoc 2007;14:478–488.
8 Ash JS, Anderson NR, Tarczy-Hornoch P: People and organizational issues in research systems implementation. J Am Med Inform Assoc 2008; 15:283–289.
9 Lee ES, et al: Incorporating collaboratory concepts into informatics in support of translational interdisciplinary biomedical research. Int J Med Inform 2009;78:10–21.
10 Lowe HJ, et al: STRIDE – an integrated standards-based translational research informatics platform. AMIA Annu Symp Proc 2009;2009:391–395.
11 Murphy SN, et al: Serving the enterprise and beyond with informatics for integrating biology and the bedside (i2b2). J Am Med Inform Assoc 2010; 17:124–130.
12 Anderson N, et al: Implementation of a deidentified federated data network for population-based cohort discovery. J Am Med Inform Assoc 2012;19: e60–e67.
13 Franklin JD, Guidry A, Brinkley JF: A partnership approach for electronic data capture in small-scale clinical trials. J Biomed Inform 2011;44:S103–S108.
14 Harris PA, et al: Research electronic data capture (REDCap) – a metadata-driven methodology and workflow process for providing translational research informatics support. J Biomed Inform 2009;42:377–381.

Mary L. Disis, MD
Center for Translational Medicine in Women's Health
850 Republican Street, 2nd Floor, Box 358050, University of Washington
Seattle, WA 98109 (USA)
E-Mail ndisis@u.washington.edu

Alving B, Dai K, Chan SHH (eds): Translational Medicine – What, Why and How: An International Perspective.
Transl Res Biomed. Basel, Karger, 2013, vol 3, pp 98–109 (DOI: 10.1159/000343144)

Engaging the Community in Research with the HealthStreet Model: National and International Perspectives

L.B. Cottler[a] · C.W. Striley[a] · C.C. O'Leary[b] · C.W. Ruktanonchai[a] · K.A. Wilhelm[c]

[a]Department of Epidemiology, College of Public Health and Health Professions and College of Medicine, University of Florida, Gainesville, Fla., and [b]Health Literacy Missouri, St. Louis, Mo., USA; [c]Faces in the Street, St. Vincent's Urban Mental Health Research Institute, Inner City Health Program, St. Vincent's Hospital, Darlinghurst, New South Wales, Australia

Abstract

We report on a model of community engagement known as HealthStreet, which began as community outreach from Washington University in St. Louis, Mo., USA, in 1989, funded by the National Institutes of Health (NIH). HealthStreet, which developed more expansive services in 2008, includes a physical site (building) which serves as a base of operation in the community from which community health workers (CHWs) can engage community members to capture real-time data on health needs, health and neighborhood concerns, and attitudes towards research through a health intake form. Based on this information, they refer community members to needed services and offer opportunities to participate in research. In November 2011, HealthStreet was opened in Gainesville (Florida, USA) as part of the community engagement efforts of the University of Florida. The HealthStreet CHWs in St. Louis and Gainesville have made contact with over 7,100 community members (83% from minorities) from January 2009 to July 2012. Recruitment and enrollment yields have been high. Participants report positive attitudes towards research and have similar health concerns and conditions across locations. HealthStreet, and other efforts that utilize CHWs to reach the community, may increase the diversity of research participants, help meet community needs, and reduce disparities in care, potentially improving public health. The international reach of the HealthStreet model is evident in the opening of HealthStreet Sydney in Australia.

In the USA alone, there are over 80,000 clinical trials conducted each year and less than 1% of the general population participates in them [1]; furthermore, nearly all of the participants are from nonminority groups. This is an issue that is global in scope: a review of 280 randomized control trials in the USA and abroad reported in six leading medical journals found that about two thirds provided no information on the race and ethnicity of their study participants [2]. Among the one third that reported information, only 13% of the participants were reported to be from minority groups. These trials also excluded women, as well as older and rural populations, leading to findings that do not account for genetic, cultural, racial, gender, age, and linguistic differences. This gap in information may then affect how well new therapies perform in the real world [3].

Even when minority populations are recruited for consideration in research studies, they are often later excluded through study criteria that limit their enrollment [4, 5]. Their exclusion is due to many factors, one of which may be that researchers and health professionals may not have the time or interest to refer or accept members of minority groups for research [6, 7]. When asthma researchers in the United Kingdom and the USA were surveyed about attitudes towards minority populations in their research, most indicated that recruiting minorities required extra time and resources; further, some believed that minority participants were unreliable [8]. However, community leaders expressed strong support for the participation of their members in research, stating a need for fairness and justice [8].

Additional challenges include limitations on transportation, time constraints due to rigid work schedules, and inappropriately low incentives offered for participation. There are also psychological factors, such as distrust of research and researchers, fears about the risk of the research to be conducted, and a general lack of understanding of the clinical research process [9]. A clinical trial that is presented to the community may not be one that reflects the concerns or interests of the community [9, 10].

Community-engaged research (CEnR), which is research that is conducted *with* community members, not *on* community members, overcomes many of these obstacles. CEnR is grounded in principles of fairness, social justice, individual empowerment, and inclusion [11], providing an opportunity for community members to form relationships with researchers based on respect and honesty [12, 13]. CEnR minimizes 'helicopter projects' conducted by investigators who 'fly' into communities, collect data, and 'fly out' when finished, leaving behind broken relationships and community mistrust.

CEnR also requires that the knowledge gained by research be shared with the participants. One way to accelerate this translation of knowledge to the end-user of health care is through one-stop shops, or front door services, where clinicians can interact directly with the public. At these locations, community health workers (CHWs) can also assess the needs and concerns of community members and then link them with relevant medical and social services, as well as with opportunities to participate in research. This effort could eventually build trust between researchers and the community.

This chapter focuses on HealthStreet, an innovative CEnR strategy that addresses all of the previously described issues. The HealthStreet model of community engagement utilizes CHWs to reach community members in the metropolitan areas of Gainesville, Florida and St. Louis, Missouri to assess health needs, health and neighborhood concerns, and attitudes towards research. Using this information, CHWs then refer members of the community to services and offer them opportunities to participate in research through the University of Florida (in Gainesville) or Washington University (in St. Louis). This chapter focuses on the methods and the types of ongoing real-time data that can be obtained from an effort such as this to contribute to the science of CEnR. It will present and compare population characteristics by study enrollment status. Its replication as an international model is addressed with a brief description of the recent establishment of HealthStreet Sydney in Australia.

Methods

The HealthStreet model began in 1989, funded by a grant from the National Institute on Drug Abuse, one of the institutes in the National Institutes of Health (NIH), to reach community members in St. Louis and link them to opportunities to participate in drug and alcohol intervention research. In May 2008, an expanded model that addressed all health problems was inaugurated as part of the community engagement activities in the NIH-funded Clinical and Translational Science Award (CTSA) that had been awarded to Washington University in St. Louis. When Dr. Linda Cottler, the Founding Director of HealthStreet St. Louis and one of the coauthors of this chapter, relocated to the University of Florida in Gainesville in 2010, she established HealthStreet in Gainesville in late 2011 as part of the community engagement efforts of the University of Florida CTSA. Now, implementation of HealthStreet Sydney is underway, and interest for other sites around the world and the USA is emerging. Components of the HealthStreet model are in practice at four other CTSA sites; they comprise the Sentinel Network for Community Engagement Project, a collaborative project of the CTSAs that facilitates community engagement across the CTSA consortium of 61 academic health centers [14].

HealthStreet is both a concept and a site. It has three main aims: to link people to services based on needs and concerns, provide opportunities to participate in health research, and build needed trust between researchers and their community. Central to those efforts are community members who are trained, coached and mentored as CHWs to deliver the model. CHWs utilize a mobile van to contact people throughout communities at bus stops, parks, grocery stores, churches, retail centers, shelters, laundromats, health fairs, and community service agencies. Importantly, CHWs engage underserved and underrepresented populations in their own neighborhoods.

After introducing themselves to community members, CHWs request consent to assess needs and concerns. Those interested provide written consent (approved by the institutional review boards of the sponsoring academic health center). Participants are then assessed with the 15-min health intake form covering demographic characteristics, attitudes towards research, history of research participation, history of medical conditions, health and neighborhood concerns, a brief drug and alcohol history, medications used currently, and access to care. Based on this information, CHWs make referrals for relevant social and medical services. Participants learn about research and about pertinent research studies. They are immediately matched to a relevant study, if desired and available, through contact with a study coordinator. They undergo detailed screening specific for the research study and sign consent forms that are specific for any study to which they are matched.

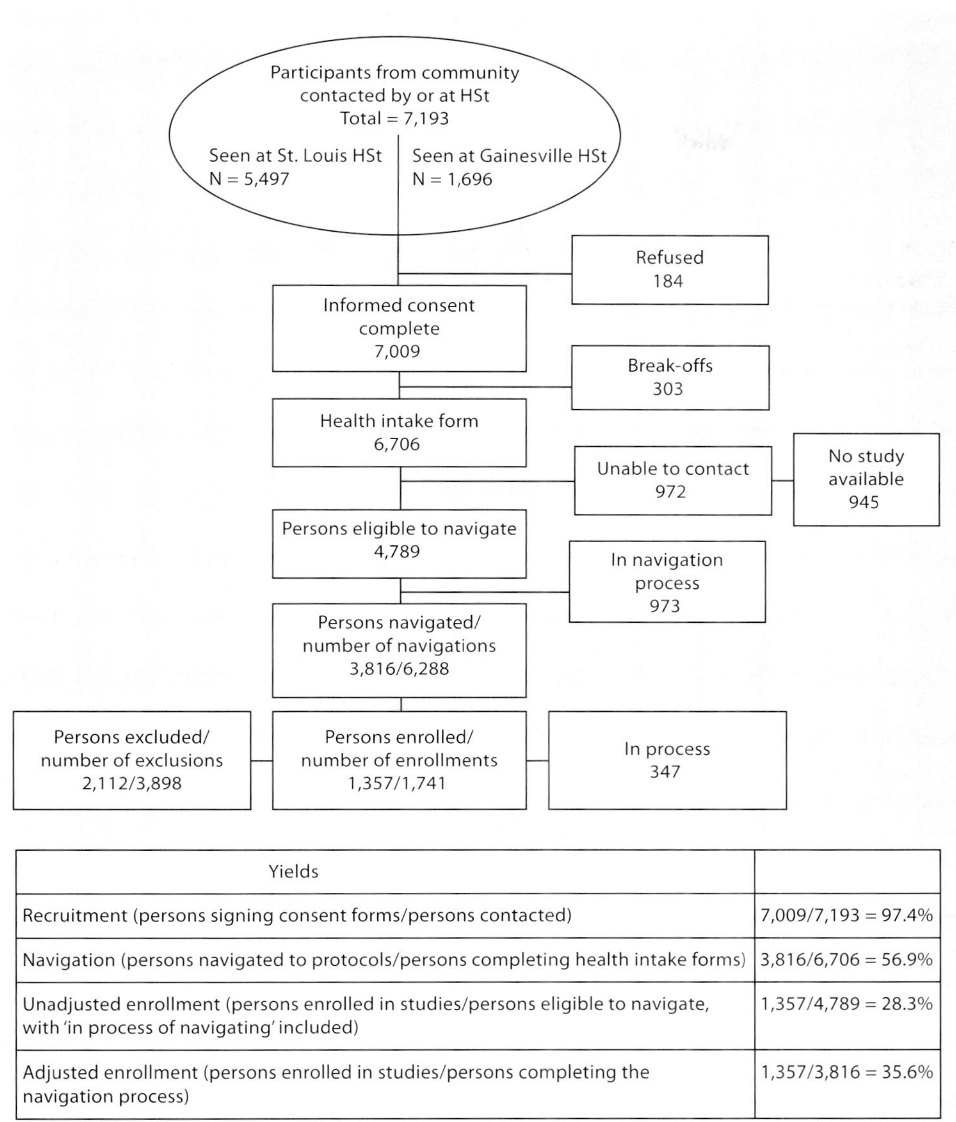

Yields	
Recruitment (persons signing consent forms/persons contacted)	7,009/7,193 = 97.4%
Navigation (persons navigated to protocols/persons completing health intake forms)	3,816/6,706 = 56.9%
Unadjusted enrollment (persons enrolled in studies/persons eligible to navigate, with 'in process of navigating' included)	1,357/4,789 = 28.3%
Adjusted enrollment (persons enrolled in studies/persons completing the navigation process)	1,357/3,816 = 35.6%

Fig. 1. Flow chart of St. Louis and Gainesville HealthStreet (HSt) participants (January 2009 through July 2012; St. Louis HSt began January 2009 and Gainesville HSt was initiated in November 2011).

Participants are later contacted by a HealthStreet 'navigator', who tracks their status regularly. The process culminates with a 30-day follow-up to determine outcomes of all referrals made, and to provide new referrals if desired. All participants are also given information about free services and classes offered at HealthStreet. In St. Louis, a Web-based system is used to enter and store information provided by community members; it also links them with applicable research studies. This Web portal is also being assessed for use in Gainesville.

Table 1. Top 5 health concerns of community recruited persons from HealthStreet in St. Louis and Gainesville by gender and age, in decades[1]

<30 years of age (n = 1,358)		30–39 years of age (n = 991)		40–49 years of age (n = 1,196)
Female	Male	Female	Male	Female
Hypertension	Diabetes	Hypertension	Hypertension	Hypertension
Weight	Hypertension	Diabetes	Diabetes	Diabetes
Diabetes	Cancer	Weight	Cancer	Weight
Asthma	Heart	Cancer	Heart	Cancer
Cancer	Asthma	Mental health	Mental health	Mental health

[1] Among those who mentioned at least one concern.

Results

Tracking Process

Because we track individuals through all activity phases, we are able to present deduplicated information on all individuals from point of contact to enrollment in specific clinical research studies (fig. 1). For the first time, we are able to present data from both HealthStreet sites. As of July 2012, CHWs have reached 7,193 individuals; 76.4% from St. Louis and 23.6% from the more recently-established HealthStreet in Gainesville. While only 2.6% of community members declined to participate, 4.2% were unable to complete the interview primarily due to having insufficient time when contacted. An additional 1,917 people were unable to be further contacted or could not be matched to a study, resulting in a total of 4,789 persons eligible to be guided or 'navigated' toward appropriate research studies. Of those, 3,816 respondents were successfully navigated to one or more research studies, resulting in a navigation yield of 56.9%. Of note, 1,357 community members were enrolled in research studies 1,741 times, yielding an unadjusted enrollment rate of 28.3%. After adjusting for navigations in process, the final adjusted enrollment rate was 35.6%.

Top Health Concerns

Table 1 reports the top five most frequently reported health concerns of community members, stratified by decade of age and gender. Hypertension and diabetes were the top two concerns for persons of all ages, with the exception of women under 30 years of age who mentioned hypertension and weight control. Men and women younger than 30 years of age differentially listed asthma in their top five concerns. Mental health began to be a concern at age 30 and persisted through age 59 for men and 49 for women. At age 60 and beyond, arthritis became one of the top five health

	50–59 years of age (n = 1,178)		≥60 years of age (n = 386)	
Male	Female	Male	Female	Male
Hypertension	Hypertension	Hypertension	Hypertension	Hypertension
Diabetes	Diabetes	Diabetes	Diabetes	Diabetes
Heart	Heart	Heart	Heart	Heart
Cancer	Weight	Cancer	Arthritis	Arthritis
Mental health	Cancer	Mental health	Cancer	Cancer

concerns. Heart disease was a concern for men 30 and older and for women 50 and older. Cancer was a top five concern for all age groups, but was no higher than third place.

Services Provided

Individuals who have been contacted through the two HealthStreet sites have already been linked to over 12,000 services, ranging from government-supported health care centers to other medical homes, food pantries, housing, testing sites for sexually transmitted diseases, eldercare, criminal justice services, dental care, and many others. CHWs in the field have routinely performed blood pressure checks; staff at the two HealthStreet sites have provided assistance to community members in completing applications for employment and have also provided nutritional counseling, among other services.

Characteristics of Enrollment

Population characteristics of HealthStreet respondents, stratified by enrollment status, are presented in table 2. Those enrolled in research studies were more likely than the nonenrolled to be male, older, and nonminority. They were also more likely to be separated, divorced or widowed, more educated, currently unemployed, and to have had recent contact with a physician.

Individuals enrolled were significantly more likely than those not enrolled to report a history of hepatitis and HIV/AIDS, depression and anxiety, hypertension, heart conditions, arthritis, and digestive problems. Respondents enrolled in studies were also more likely to report recent marijuana use as compared to those who were not enrolled.

One of the most interesting aspects of the study is the ongoing assessment of community attitudes and perceptions about research (table 3). As shown, 23.8% of those enrolled in studies reported prior participation in research, compared to 18.3% of those not enrolled. Enrolled and nonenrolled individuals had similar responses concerning research studies: nearly all were interested in participating in studies, including those in which blood samples would be taken and also those in which genetic testing would be done. The lowest enthusiasm was for studies in which participants would have to take medication and also for those which did not provide any financial compensation. On average, nonenrolled respondents reported desiring higher compensation for research.

Discussion

With a required focus on community engagement in the CTSA program, which was initiated by the NIH in 2006, each academic health center had to find ways to foster new relationships and partnerships with the community. While many academic health centers focus primarily on practice-based networks or partner with federal, state, or nonprofit agencies that provide support to the community, our model fosters approaches that actually seeks the participation of community members.

Protocols and strategies to recruit research participants who are not in a health care system or receiving heath care from a professional are being developed sporadically across the world. One approach utilizes registries, which require individuals to log on to a website and provide their lifetime medical history; this is done without any contact with staff. This passive strategy may be ideal for highly-motivated volunteers with specific research interests; however, this passive method does not diversify the population base. Thus, results based on registries may not be generalizable, and findings will be difficult to translate into routine clinical practice.

HealthStreet is one CEnR model which is reaching people where they live, translating findings into easily understood terms and concepts, building trust, and increasing opportunities for participation in research studies. HealthStreet has been effective in linking underrepresented populations to health care in two geographic areas – St. Louis and Gainesville. Data collected from over 7,100 participants support this innovative method to deliver health messages and link people to services and opportunities to participate in health research. The community contacted has been diverse in every way, from location to race, age, and research interests.

Community members have been referred to a variety of services through HealthStreet, and report general satisfaction with those referrals. We have just begun to assess the outcomes and satisfaction with these services in a more rigorous fashion. We are using feedback from the community members to seek additional services and educational and research opportunities to meet their expressed needs.

Reaching populations that are predominantly minority and reducing obstacles to their participation in clinical research have been two of the major outcomes of the program. We have found, based on data from over 6,000 participants seen through

Table 2. Characteristics of enrolled vs. not enrolled community-recruited persons from HealthStreet through July 2012[1]

Demographics	Enrolled (n = 1,357), %	Not enrolled (n = 2,112), %	p
Female	53.7	58.4	0.007
Mean age	42.7 years (±13.4)	38.5 (±13.7)	<0.0001
Ethnicity/race reported			
Asian	0.5	0.2	<0.0001
African American	76.5	83.6	
Caucasian	18.8	14.4	
Other	4.2	1.8	
Minority	81.4	85.7	0.0007
Hispanic/Latino	1.4	2.0	0.18
Marital status			
Never married	59.9	66.2	<0.0001
Married	12.2	13.1	
Separated/divorced/widowed	27.9	20.7	
Education			
< High school	22.0	25.0	0.0001
High school	38.9	43.0	
> High school	39.1	32.0	
Employed	26.1	33.9	<0.0001
Seen a doctor in past 6 months	65.0	60.8	0.02
Had health check-up in past 12 months	69.3	66.9	0.20
Any insurance	47.1	52.8	0.003
Sexually transmitted diseases			
Chlamydia	7.4	6.2	0.21
Gonorrhea	4.7	5.1	0.66
Hepatitis	5.0	3.2	0.02
Herpes	2.5	1.7	0.13
HIV/AIDS	3.0	1.9	0.04
Syphilis	0.6	0.8	0.82
Any STD above	16.7	12.2	0.0002
Mental health disorders			
Affective disorder	35.3	24.8	<0.0001
Anxiety	17.9	12.7	<0.0001
ADD/ADHD	5.1	3.9	0.11
Autism	0.3	0.3	0.66
Eating disorder	1.8	1.7	0.73
Personality disorder	1.5	1.3	0.66
Schizophrenia	4.9	0.8	0.06
Gambling problem	0.6	4.2	0.04
Any mental health disorder	39.4	29.1	<0.0001
Medical disorders			
Hypertension	33.1	29.1	0.01
Diabetes[1]	10.8	10.3	0.67
Asthma[1]	17.8	19.1	0.36

Table 2. continued

Demographics	Enrolled (n = 1,357), %	Not enrolled (n = 2,112), %	p
Any cancer'1'	9.2	9.3	0.97
Weight problem/obesity	13.8	13.3	0.63
Heart problem'1'	3.1	1.9	0.03
Arthritis	24.4	20.3	0.005
Digestive problem	26.6	19.1	<0.0001
Neurological problems'1'	8.2	6.9	0.14
Any medical disorder above with '1'	33.5	31.0	0.13
Recent (past 30 days) substance use			
Tobacco	43.0	43.3	0.96
Marijuana'2'	27.3	16.7	0.04
Cocaine'2'	1.8	2.5	0.68
Opioids	15.2	12.5	0.53
Any drugs above with '1'	37.0	25.8	0.05

STD = Sexually transmitted disease; ADD/ADHD = attention deficit disorder/Attention deficit hyperactivity disorder.

'1' Data are incomplete for some cells.

the CTSA Sentinel Network, that minority populations have recently reported favorable attitudes towards research participation [14]. We attribute this willingness to the CHWs, who engage each participant with warmth and careful listening.

An early but persistent finding has been a disconnect between high interest in participation and low availability of studies, suggesting clinical research has not kept pace with the needs and wants of the community. Additionally, typical remunerations have been lower than what the community members desire. While it is possible that such disconnects could perpetuate barriers to research participation seen in the past, our data and experiences suggest that a team that is persistent and engaging can minimize these barriers, but should remain vigilant to the reappearance of obstacles.

The community engagement field is emerging and strives not only to provide a voice to all communities, but also to measure outcomes; metrics have been developed to track in real-time ongoing contacts, services, and outcomes with the intention of advancing the science of CEnR not just in the USA, but around the world.

Recently, investigators in Paris, France and Lima, Peru have expressed interest in the HealthStreet model after learning about the imminent opening of HealthStreet Sydney in Australia, and it will be interesting to see how the HealthStreet model performs internationally, and in different settings. The Sydney HealthStreet will be part of the Inner City Health Program at St. Vincent's Hospital in the inner city; it will be the hub of an integrated network of health and social services, offering a one-stop street office and a mobile van to increase access to relevant, effective services and to

Table 3. Research perceptions of HealthStreet respondents by enrollment status[1]

	Enrolled (n = 1,357)		Not enrolled (n = 2,112)	
	n	%	n	%
Have you ever been in a health research study?				
No/don't know	950	76.2	1,571	81.7
Yes	297	23.8	351	18.3
Would you volunteer for a health research study:				
That only asked questions about your health?				
No	95	10.7	98	9.8
Yes	793	89.3	898	90.2
If research wanted to see your medical records?				
No	68	7.7	65	6.5
Yes	819	92.3	930	93.5
If you had to give a blood sample?				
No	53	6.0	41	4.1
Yes	834	94.0	955	95.9
If you were asked to give a sample for genetic studies?				
No	68	7.7	83	8.3
Yes	819	92.3	912	91.7
If you might have to take medicine?				
No	231	26.2	303	30.5
Yes	652	73.8	690	69.5
If you were asked to stay overnight in a hospital or clinic?				
No	109	12.3	144	14.5
Yes	776	87.7	850	85.5
If you might have to use medical equipment?				
No	81	9.2	80	8.0
Yes	802	90.8	915	92.0
Would you participate in a study if you didn't get paid?				
No	218	27.5	291	32.6
Yes	576	72.5	602	67.4
How interested are you in being in a research study?				
Not at All	10	1.3	6	0.7
Maybe	160	20.1	196	22.0
Definitely	626	78.6	690	77.4
How much money do you think is a fair amount for participation in a study that lasts about an hour and a half and involves an interview and a blood test?	USD 72.56 (±117.67)		USD 84.62 (±132.12)	

[1] Data are incomplete for some cells.

improve the quality of that care. HealthStreet Sydney will facilitate access to health care through CHW- or nurse-provided screenings for hypertension and diabetes and navigate people to appropriate services, just as the original HealthStreet model is doing. It is also utilizing an e-health authoring platform that is an initiative from the Medical School at the University of New South Wales, funded by the Department of Health Australia. It is designed to promote access to the main databases, providing information about drug and alcohol services, accommodation, and programs in Sydney, as well as mental health materials and other Web-based interventions (http://www.mindhealthconnect.org.au/).

A recent study of stakeholders in Sydney reported a concern for identifying people who needed services before they reached a crisis point. That study and also a health impact assessment for HealthStreet Sydney conducted by the Center for Primary Health Care and Equity, University of New South Wales, have formed the springboard for the Sydney project [15]. One conclusion was that the success of HealthStreet Sydney will be determined by the relationships developed between and among the health and social service organizations, a primary mission of the HealthStreet model.

Engaging the community in research with the HealthStreet model serves multiple purposes locally, nationally, and internationally. It builds a cohort of community residents who are available for research opportunities, offers a voice to those who may not have a say in what medical and social services are needed, becomes a centralized information hub where findings can be translated, and is a place where community residents and researchers can exchange information. This innovative HealthStreet model thus produces outcomes that matter to improve the health of communities, both nationally and internationally.

Acknowledgements

We would like to acknowledge community member participants from both HealthStreet locations and St. Louis Coordinator Erin Murdock and Gainesville Coordinator Noni Graham. This work is supported in part by the Washington University Institute of Clinical and Translational Sciences grant UL1 TR000448 and the University of Florida College of Public Health and Health Professions and College of Medicine Clinical and Translational Science Award grant UL1 TR000064 from the National Center for Advancing Translational Sciences of the NIH, in addition to Community Engagement and Sentinel Network to Increase Community Participation in Research, grant UL 1RR029890. The project is also supported by RC2 HL101838 from the National Heart, Lung and Blood Institute of the NIH. The content is solely the responsibility of the authors and does not necessarily represent the official view of the NIH.

References

1 Mozes A: Report claims clinical trials miss many populations. HealthDay News April 1, 2008.

2 Rochon PA, Mashari A, Cohen A, Misra A, Laxer D, Streiner DL: The inclusion of minority groups in clinical trials: problems of under representation and under reporting of data. Account Res 2004;11: 215–223.

3 Murthy VH, Krumholz HM, Gross CP: Participation in cancer clinical trials: race-, sex-, and age-based disparities. JAMA 2004;291:2720–2726.

4 Rothwell PM: External validity of randomized controlled trials: 'to whom do the results of this trial apply?'. Lancet 2005;365:82–93.

5 Charleson ME, Horwitz RI: Applying results of randomized trials to clinical practice: impact of losses before randomization. Br Med J 1984;289: 1281–1284.

6 Ford JG, Howerton MW, Bolen S, Gary TL, Lai GY, Tilburt J, Gibbons MC, Baffi C, Wilson RF, Feuerstein CJ, Tanpitukpongse P, Powe NR, Bass EB: Knowledge and access to information on recruitment of underrepresented populations to cancer clinical trials. Evid Rep Technol Assess (Summ) 2005;122:1–11.

7 Douglas A, Bhopal RS, Bhopal R, Forbes JF, Gill JMR, Lawton J, McKnight J, Murray G, Sattar N, Sharma A, Tuomilehto J, Wallia S, Wild SH, Sheikh A: Recruiting South Asians to a lifestyle intervention trial: experiences and lessons from PODOSA (Prevention of Diabetes & Obesity in South Asians). Trials 2011;12:220.

8 Sheikh A, Halani L, Bhopal R, Netuveli G, Partridge MR, Car J, Griffiths C, Levy M: Facilitating the recruitment of minority ethnic people into research: qualitative case study of South Asians and asthma. PLoS Med 2009;6:e1000148.

9 Striley CW, Callahan C, Cottler LB: Enrolling, retaining and benefitting out-of-treatment drug users in intervention research. J Empir Res Hum Res Ethics 2008;3:19–25.

10 Williams MM, Scharff DP, Mathews KJ, Hoffsuemmer JS, Jackson P, Morris JC, Edwards DF: Barriers and facilitators of African American participation in Alzheimer's disease biomarker research. Alzheimer Dis Assoc Disord 2010;24:S24–S29.

11 Centers for Disease Control and Prevention (CDC): Principles of Community Engagement, ed 2. Atlanta, CDC/ATSDR Committee on Community Engagement, 2011.

12 Adderley-Kelly B, Green P: Strategies for successful conduct of research with low income African American populations. Nurs Outlook 2005;53:147–152.

13 Allman RM, Sawyer P, Crowther M, Strothers HS, Turner T, Fouad MN: Predictors of a 4-year retention among African American and white community-dwelling participants in the UAB study of aging. Gerontologist 2011;51:S46–S58.

14 Cottler LB, McCloskey DJ, Agiular-Gaxiola S, Bennett NM, Dwyer-White M, Collyar DE, Ajinka S, Seifer SD, O'Leary CC, Striley CW, Evanoff B: Geographic and racial/ethnic differences in community needs, concerns and perceptions about health research: findings from the CTSA Sentinel Network. Am J Pub Health; in press.

15 Wilhelm K, Wise M, Haigh F, Collins L, Bernardi S, McGeorge P: Connecting the dots, finding the gaps: solutions provided by the HealthStreet project (submitted for publication).

Linda B. Cottler, PhD, MPH
Dean's Professor and Chair
Department of Epidemiology, University of Florida
College of Public Health and Health Professions, College of Medicine
1225 Center Drive, PO Box 100231, Gainesville, FL 32611 (USA)
E-Mail Lbcottler@UFL.edu

Alving B, Dai K, Chan SHH (eds): Translational Medicine – What, Why and How: An International Perspective.
Transl Res Biomed. Basel, Karger, 2013, vol 3, pp 110–119 (DOI: 10.1159/000343025)

Evaluating Translational Research

Cathleen Kane[a] · Doris Rubio[b] · William Trochim[a]

[a]Cornell University, Ithaca, N.Y., and [b]University of Pittsburgh, Pittsburg, Pa., USA

Abstract

Evaluators face challenges such as program and intervention variability, time and budgetary constraints, and a complicated range of internal and external stakeholders. The design and implementation of evaluation for translational research (TR) is especially challenging since there is no universal agreement about what exactly 'translation' means. A range of effective methodologies for the evaluation of TR that have been used in the context of the Clinical and Translational Science Award (CTSA) program include: process analysis, research publication analysis, cost analysis, surveys and interviews, qualitative methods, and experimental and quasi-experimental designs. Examples of each are described in this chapter. The methods used to evaluate TR constitute their own independent field of research, representing the formal study of how to encourage scientific research-to-practice or the 'science of science'. As such, the evaluation of TR and the broader field of medical research have a symbiotic relationship that will be of increasing importance as both evolve to meet future challenges and opportunities.

Evaluation is the systematic acquisition and assessment of information to provide useful feedback about some object [1]. Here, the object could be the process of translational research (TR), an activity associated with translation, or an attempt to improve translation or remove barriers. Unlike other more controlled forms of research, evaluation almost invariably occurs in the everyday messy context of programs and interventions, has to be conducted within considerable constraints, and provides results and conclusions to stakeholder groups who may have competing and often conflicting views. It is used for a variety of purposes including the monitoring and improvement of programs, the provision of data and analyses for management decision-making, and the generation of knowledge about how a program or intervention works and whether it is effective.

One of the major challenges for evaluating TR is the elusiveness of the definition of translation itself [2]. There is no universal agreement about what translation means. Some suggest translation should be confined to the 'connection between basic and clinical research' [3], while others argue that it covers a much broader range of the continuum from basic research to health impacts [4–6]. There is considerable variability in the description of what is being translated [4, 7–14], or where translation starts or ends [8, 9, 13, 15–27]. Finally, there are a variety of proposed systems for classifying the phases of TR that range from T2 ('two phase') systems [28] to T3 or T4 models [29–31]. These definitional and conceptual varieties pose significant challenges for evaluation and have led to efforts to develop operationally definable and measureable process 'marker' models of TR [2] that enable evaluation to take place within and be consistent across a broad range of alternative multiphase definitional frameworks [32].

Types of Translational Research Evaluation

The term 'evaluation' can be applied to a bafflingly broad set of activities, all of which have relevance to TR. For instance, the area of tracking and monitoring and the associated use of one or more metrics for measuring progress can be considered a type of 'monitoring evaluation', most typically used for program management and accountability purposes. In contrast, a controlled experimental study of an intervention designed to enhance collaboration among researchers could also be considered a form of evaluation, in this case, 'evaluation outcome research'. 'Prospective' (or 'ex ante') evaluation is typically distinguished from 'retrospective' (or 'ex post') evaluation. The former involves a broad range of activities, including conceptualization and program model development, needs assessment, and evaluability assessment. The latter includes the bulk of what is commonly viewed as evaluation and includes monitoring and assessment. 'Process' (or 'formative') evaluation examines how a phenomenon unfolds or how an intervention is sequenced and is distinct from 'outcome' (or 'summative') evaluation that looks at whether an intervention or event caused outcomes to occur.

Evaluation of TR, as well as other forms of research, can be classified broadly as observational, correlational, or causal in intent. Most monitoring is observational in nature. Correlational evaluation examines factors that are associated with translational progress. Causal evaluation typically involves designing and conducting a controlled trial of the effects of an intervention to affect TR.

Evaluation can be conducted by evaluators who are internal or external to a program. Often internal evaluation is done locally within an organization to obtain rapid feedback and to monitor ongoing processes. External evaluation, which is conducted by evaluators who have no direct association with the program or organization that manages it, is usually conducted at the request of a funding agency to ensure accountability and to determine program effectiveness.

We often distinguish between 'quantitative' and 'qualitative' evaluations based on the type of measurement approaches that are used. Most contemporary evaluations tend to utilize a 'mixed methods' evaluation approach [33] that combines or integrates both types of measures.

The major variables of interest in TR evaluation can be divided into two categories: core variables and moderator variables. Core variables might best be described with the familiar phrase: 'better, faster, cheaper'. One of the central aims of TR is to reduce the length of time for medical innovations (e.g. drug development, new devices, treatments) to be applied (faster) while at least keeping cost (cheaper) and quality (better) constant [29, 34]. The moderating variables in TR are factors that potentially influence the core variables, including but not limited to: the effectiveness of collaboration of researchers; training of clinicians, researchers, and practitioners; and improvements in institutional support, policies, and infrastructure.

Methods for Evaluating Translational Research

Some of the most promising TR evaluation approaches can be illustrated with examples of methodologies used in Clinical and Translational Science Award (CTSA) evaluation contexts [35].

Process Analysis
A recognized barrier in clinical research is the length of time required for a protocol to be developed and then approved by the institutional review board (IRB) [36, 37]. Multiple CTSA institutions worked together to develop and implement a study that measured key process marker points for research protocols that were successfully submitted to the IRB at participating CTSA sites. These duration metrics included the elapsed number of days between key process marker dates including date of protocol submission, return to investigators for prereview revision, return by investigators with revisions, full IRB committee review, return to investigators for postreview revision, submission of revised protocol, and date of final approval. The purpose of the study was to develop objective metrics to be used to implement, monitor, and standardize process improvement within and across the CTSA consortium. Additional evaluation has focused on the process of accrual of subjects into clinical trials [38], and the inclusion of research publications into meta-analysis, such as the Cochrane Collaboration. Figure 1 shows a hypothetical translational continuum with major process 'markers' [2], and illustrates how one can 'drill down' to study more detailed sub processes, such as illustrated in the IRB study.

Research Publication Analysis
Bibliometrics and publication analysis are widely recognized methods for measuring research performance [39]. While publication data are themselves often considered short-term 'outcomes' in TR, by using coauthorship and citation data, the effects of the key 'process' or 'moderating' variable of interdisciplinary collaboration can also be assessed. One important resource for evaluators is the use of benchmarked publication databases such as PubMed and the Web of Science, which allow for in-depth analysis via comparison to historical norms and established metrics. With these databases, evaluators have measured not just publication rates, but also the quality of research publications as reflected in citations when compared with all other

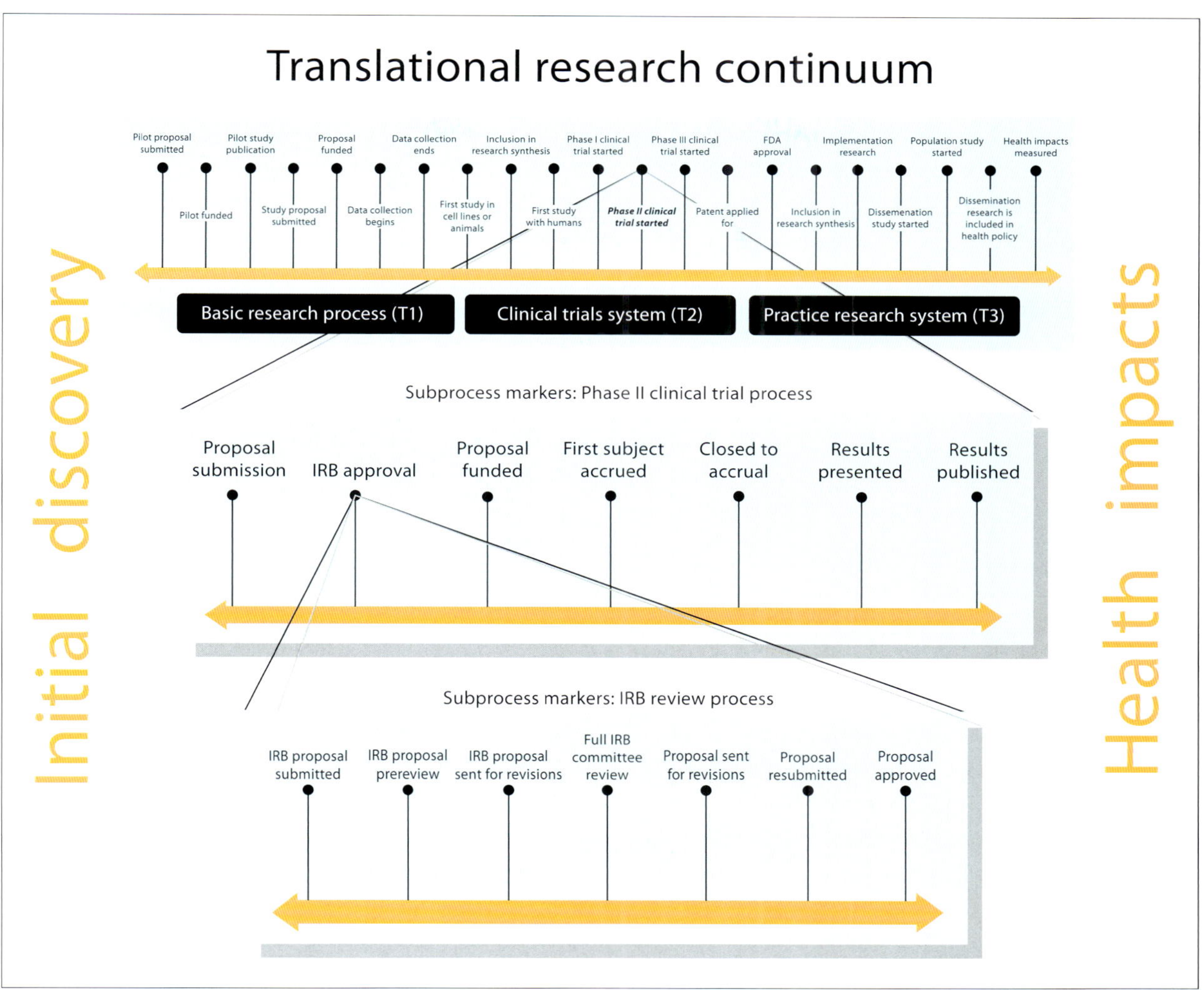

Fig. 1. A process marker model of the TR continuum showing examples of subprocesses (adapted from Trochim et al. [2]).

publications of the same type in the same journals or fields over a specified window of time. There are a wide number of metrics that have been developed using publication data [40], including:

- Number of authors/average number of authors per paper: these metrics include the total number of unique author names found in the papers in the journal or in the field under analysis as well as the average number of authors per paper, representing the extent of collaboration
- Disciplinarity index: a measure that reflects the level of multidisciplinarity in a set of papers; ranges from 0–1, with lower numbers reflecting greater multidisciplinarity

- Citations and the C-index: equals the ratio of actual citations to expected citations for a group of publications, where the comparison rate of expected citations is based on all articles of the same type (e.g. research article) for its journal or field and year (indexed year)

TR evaluators have used metrics like these to illustrate both the quantity (number of citations) and the quality (use by peers) of research productivity. These bibliometrics are also useful in measuring TR collaboration from two parallel perspectives: size and discipline. With the basis for comparison provided by the Web of Science, a CTSA institution can determine if its research teams are inclusive (e.g. number of authors) and if its center has succeeded in encouraging a better than average number of collaborations across areas of expertise (e.g. disciplinarity index). Bibliometrics are one of the few areas where TR evaluators have access to true comparative benchmarks and historical data.

Cost Analysis

Return on investment (ROI) is a commonly used cost analysis approach for assessing the financial consequences of investments or business decisions. Most simple ROI analysis divides the incremental economic gain from an action by the investment costs, resulting in an ROI ratio. When controlled for similar circumstances, the higher the ROI, the greater the financial return and, all things being equal, the better the decision or investment. Salaries and wages are considered direct costs, and can be attributed directly to the investment (e.g. a CTSA-funded TR research protocol). Proximal measures of gains, or returns, are also included, insofar as they can be tracked for a sufficient length of time and tied directly to the specific investment (e.g. a CTSA-funded TR research protocol subsequently competes successfully for additional federal funding). This kind of analysis can be useful, but grows in complexity with the recognition of several important dimensions of the economic value implied by the ratio. A good example of this type of cost analysis is a study of the projected spillover effects of a CTSA into the local economy, conducted at the University of Rochester Medical Center (URMC) [41]. URMC was selected to be one of the first CTSAs, and as part of their start-up phase, center leadership postulated that this substantial National Institutes of Health (NIH) grant would not only affect URMC's ability to support TR, but that it would also reinforce their contribution to the regional economy. In a formal study, URMC used ROI to estimate the direct, indirect, and induced economic impact of the CTSA grant, as well as all the other expected investments directly tied to the grant funding. The analysis indicated that the local CTSA project's immediate and catalytic impacts would total about USD 43 million in labor income and over 600 jobs.

Surveys and Interviews

Surveys and interviews are efficient ways to gather systematic data on numerous factors, such as an expressed need for (e.g. needs assessment) or satisfaction with services. The methodology is flexible in that the surveys can be administered using the internet, telephone, face-to-face contact, or mail. Many evaluators within the CTSA program utilize surveys and interviews to gather information about the investigators whom they serve and the trainees of their center. For example, if a CTSA works with several hundred investigators over the course of one year, the evaluator would want to survey all of the investigators (or a representative sample) to assess their satisfaction, gather success stories, and collect suggestions for improvement regarding the support services provided.

One example of how CTSA evaluators have used surveys or interviews is in the assessment of trainees. Utilizing a logic model approach that looks at such factors as financial input into the program, training courses, and outcomes (short-term and long-term), Rubio et al. [32] developed a model of how to assess training in TR and how such training differs from training in clinical research or basic science. TR training involves teachers from numerous disciplines to provide the breadth of information that enables the trainee to understand what basic science and clinical research involve, as well as how this information gets translated into clinical practice. At the University of Pittsburgh, a team of investigative evaluators created a model of career success for clinical and translational researchers [42]. The evaluators conduct annual surveys of the trainees to assess numerous factors that contribute to career success. Finally, an education workgroup of the CTSAs adopted the University of Pittsburgh's model and developed numerous metrics and measures in concert with the model to assess the career trajectories of trainees [43]. All of these initiatives advocate the use of survey methodology in different ways to assess the career success of trainees.

Experimental and Quasi-Experimental Methods

Experimental designs are particularly strong at assessing cause-and-effect relationships [44]. These types of designs are recommended for established programs or interventions with available control or comparison groups that allow random assignment of participants. As such, TR evaluators must assess the stability of the program or intervention in question before using such approaches; statisticians are needed to oversee the design, randomization, and analyses of the study. If the intervention or program is still in an initial more dynamic stage of development, these designs are not ideal [45].

Experimental and quasi-experimental designs (as well as the use of multivariate and regression analysis) are emerging as some of the most promising new collaborative evaluation approaches within the CTSAs. In 2011, for example, sixteen CTSA academic medical centers conducted a multisite randomized block controlled trial [46] to evaluate whether a structured research mentoring curriculum would measurably improve the mentoring skills of research mentors. The mentors assigned to the intervention (the structured curriculum) showed significantly improved mentoring practices relative to control mentors. The consensus among the participating TR evaluators and program stakeholders was that such approaches hold promise both for improving mentoring interventions and as an evaluation methodology across institutions with CTSAs.

Qualitative Methods

Qualitative data can help evaluators learn about the specific set of motives, incentives, and contextual details that move the germ of an idea from its inception as a scientific hypothesis to its application as a concrete clinical practice (health impacts). Given the proper rigor and design, qualitative evaluative methods, such as case studies, interviews and focus groups, can greatly benefit TR evaluation. Mixed methods research combines qualitative and quantitative methods and capitalizes on the strengths of both. For evaluators taking on the numerous and complex variables associated with TR, qualitative data can literally help quantitative data 'come alive' by illustrating the various pathways on the TR continuum.

For example, the Weill Cornell CTSA is developing case studies that can serve as essential evaluation components, with a 'case' defined as a drug, medical device, or a surgical process/procedure that has already been translated into clinical practice. FDA approval, Cochrane Reports, and Medicare/Medicaid approval serve as important marker points in successful translation. Once these cases and markers have been identified, evaluators engage in an activity that could be described as 'translational forensics', in which they use publically available data sources such as literature reviews, patent records, and data mining in order to determine a genesis point for the cases. This information is then used to identify key participants, who are interviewed to document the entire story of the research translation from the initial discovery point(s) to clinical application(s). The interview data are analyzed and collated, relevant durations are calculated, and the TR pathways and timelines are depicted visually.

The Future of Translational Research Evaluation

TR cuts across the entire research endeavor from discovery to impact. It includes all of the traditional phases of research (basic, clinical, applied, etc.), is dynamic rather than simple, is bidirectional (e.g. can move from clinical back to basic), and involves multiple participants throughout the research endeavor. In an increasing sense, everyone who is involved anywhere along the TR continuum can be considered a participant in the process of translation. This has profound implications for the scope of work that TR evaluation is expected to cover.

The CTSA initiative of the US NIH is one of the major investments to date in the infrastructure and support of TR. It is likely that many of the trends in TR and its evaluation will emerge from that consortium. For instance, there have been suggestions that the CTSA consortium could function as a 'virtual national laboratory for reengineering clinical translational science' [47], with multiple sites collaborating on common evaluation protocols to assess the effectiveness of different interventions for enhancing translation. The CTSA Evaluation Key Function Committee, consisting of representatives from all 61 site evaluation teams has developed evaluation guidelines [48] that provide high-level policy guidance for the evaluation of TR.

In specific substantive areas of TR, CTSA groups have advanced TR evaluation. For instance, the Biostatistics, Epidemiology and Research Design (BERD) Key Function Committee has developed a set of metrics [49], as has the Education Key Function Committee [42], and the Community Engagement Key Function Committee has developed general approaches for that particularly challenging context for evaluation [50]. However, the CTSAs are not the only entities working on evaluating TR. For instance, in the HIV/AIDS clinical research networks sponsored by the NIH, considerable work has been done on the development of a clinical research management information system that enables more advanced evaluation of the process of starting and running a clinical trial [36]. Furthermore, the professional field of evaluation has been working on TR as a science management endeavor through its Topical Interest Groups on Health Evaluation, and Research, Technology and Development Evaluation, among others. All of these forces are actively shaping the future of TR evaluation.

TR is increasingly thought to encompass both TR and 'research on translation'. In some sense, there is no particular research that cannot be characterized as translational – *all* research is moving through the translational continuum. The evaluation of TR is the study of the process of how translation happens, what affects it, and how it can function better. In this sense of its meaning, TR is the 'science of science', the formal study of how to encourage research-to-practice. This notion is already emerging in multiple subconstructs such as 'science of science policy' [51] and 'science of team science' [52, 53]. Evaluation is its metamethodology, providing leaders and managers with the data they need to make informed decisions about how to improve or influence translation.

Acknowledgements

This project has been funded in whole or in part with Federal funds from the National Center for Research Resources and National Center for Advancing Translational Sciences (NCATS), National Institutes of Health (NIH), through the Clinical and Translational Science Awards Program (CTSA) by grants UL1 RR024996 and UL1 TR000457 (Weill Cornell); and UL1 RR024153 and UL1 TR000005 (University of Pittsburgh).

References

1 Trochim W, Donnelly JP: The Research Methods Knowledge Base, ed 3. Cincinnati, Thomson (Atomic Dog) Publishing, 2006.

2 Trochim W, et al: Evaluating translational research: a process marker model. Clin Transl Sci 2011;4: 153–162.

3 Cesario A, et al: The role of the surgeon in translational research. Lancet 2003;362:1082.

4 Devieux JG: Cultural adaptation in translational research: field experiences. J Urban Health 2005;82: iii82–iii91.

5 Baum C: Transition. Occupation Ther J Res 2004; 24:43.

6 Birmingham K: What is translational research? Nat Med 2002;8:647.

7 Fain JA: Making the case for translational research. Diabetes Educ 2004;30:162.
8 Dowsett M: Translational research and the changing face of breast cancer. Breast Cancer Res Treat 2004;87(Suppl 1):S1–S2.
9 Kim R: Internet-based Profiler system as integrative framework to support translational research. BMC Bioinformatics 2005;6:304.
10 Liang MH: Translational research: getting the word and the meaning right. Arthritis Rheum 2003;49: 720–721.
11 Olatunji BO: Graduate training of the scientist-practitioner: issues in translational research and statistical analysis. Behav Ther 2004;27:45–50.
12 Stahl A: Positron emission tomography as a tool for translational research in oncology. Mol Imaging Biol 2004;6:214–224.
13 Watts R: Translational research in autoimmunity: aims of therapy in vasculitis. Rheumatology 2005; 44:573–576.
14 Corrigan PW: Demonstrating translational research for mental health services: an example from stigma research. Ment Health Serv Res 2003;5:79–88.
15 Fontanarosa PB, DeAngelis CD: Basic science and translational research – call for papers. JAMA 2001; 285:2246–2246.
16 Richards JS: Spinal cord injury pain: impact, classification, treatment trends, and implications from translational research (miscellaneous). Rehab Psychol 2005;50:99–102.
17 Nunes EV, Carroll KM, Bickel WK: Clinical and translational research: introduction to the special issue. Exp Clin Psychopharmacol 2002;10:155–158.
18 Hightower LE: Introducing Professor Stuart Calderwood, Stress Response Translational Research section editor. Cell Stress Chaperones 2004;9:1–3.
19 Apolone G: Controversial effect of epoetin in cancer: grounds for a translational research exercise? Eur J Cancer 2004;40:1289–1291.
20 Crist TB, Schafer AI, Walsh RA: Translating basic discoveries into better health care: the APM's recommendations for improving translational research. Am J Med 2004;116:431–434.
21 Pazdur R: Cancer drug development and translational research are high priorities of special workshop. Cancer Biol Ther 2004;3:703–704.
22 Sugarman J, McKenna WG: Ethical hurdles for translational research. Radiat Res 2003;160:1–4.
23 Ter Linde JJM, Samsom M: Potential of genetic translational research in gastroenterology. Scand J Gastroenterol 2004;39:38–44.
24 Clemens JD, Jodar L: Translational research to assist policy decisions about introducing new vaccines in developing countries. J Health Popul Nutr 2004; 22:223–231.
25 Garfield SA, et al: Considerations for diabetes translational research in real-world settings. Diabetes Care 2003;26:2670–2674.
26 Licinio J, Wong ML: Translational research in psychiatry: pitfalls and opportunities for career development. Mol Psychiatry 2004;9:117.
27 Buckley NA, Eddleston M, Dawson AH: The need for translational research on antidotes for pesticide poisoning. Clin Exp Pharmacol Physiol 2005;32: 999–1005.
28 Sung NS, et al: Central challenges facing the national clinical research enterprise. JAMA 2003; 289:1278–1287.
29 Westfall JM, Mold J, Fagnan L: Practice-based research – 'Blue Highways' on the NIH roadmap. JAMA 2007;297:403–406.
30 Dougherty D, Conway PH: The '3T's' road map to transform US health care. JAMA 2008;299: 2319–2321.
31 Khoury MJ, et al: The continuum of translation research in genomic medicine: how can we accelerate the appropriate integration of human genome discoveries into health care and disease prevention? Genet Med 2007;9:665–674.
32 Rubio DM, et al: Defining translational research: implications for training. Acad Med 2010;85: 470–475.
33 Greene JC, Caracelli VJ, Graham WF: Toward a conceptual framework for mixed-method evaluation designs. Educ Eval Policy Anal 1989;11: 255–274.
34 Balas EA, Boren SA: Managing clinical knowledge for healthcare improvement; in: Yearbook of Medical Informatics. Stuttgart, Schattauer Verlagsgesellschaft mbH, 2000.
35 Creswell JW, et al: Best Practices for Mixed Methods Research in the Health Sciences. Bethesda, NIH/OBSSR, 2011.
36 Kagan JM, Rosas SR, Trochim W: Integrating utilization-focused evaluation with business process modeling for clinical research improvement. Res Eval 2010;19:239–250.
37 Drezner MK, Cobb N: Efficiency of the IRB review process at CTSA-sites; in: CTSA Clinical Research Management Workshop. New Haven, NIH/NCATS/CTSA, 2012. www.ctsacentral.org.
38 Kitterman DR, et al: The prevalence and economic impact of low-enrolling clinical studies at an academic medical center. Acad Med 2011;86:1360–1366.
39 Rosas SR, et al: Evaluating research and impact: a bibliometric analysis of research by the NIH/NIAID HIV/AIDS clinical trial networks. PLoS ONE 2011; 6:e17428.

40 Thomson Reuters, Using Bibliometrics: A Guide to Evaluating Research Performance with Citation Data. Philadelphia, Thomson Reuters, 2008.
41 Gardner K: Impact of Award and Spillover Effects. Rochester, University of Rochester Medical Center CTSA, 2007. www.policyarchive.org/handle/10207/bitstreams/11078.pdf.
42 Lee LS, et al: Clinical and translational scientist career success: metrics for evaluation. Clin Transl Sci, 2012, in review.
43 Rubio DM, et al: A comprehensive career-success model for physician-scientists. Acad Med 2011;86:1571–1576.
44 Cook TD, Campbell DT: Quasi-Experimentation: Design and Analysis for Field Settings. Boston, Houghton Mifflin Company, 1979.
45 Urban JB, Trochim W: The role of evaluation in research-practice integration: working toward the 'golden spike'. Am J Eval 2009;30:538–553.
46 Pfund C: Training Mentors of Clinical and Translational Research Scholars: A Randomized Controlled Trial (under review). Madison, University of Wisconsin, 2012.
47 Dilts D, Rosenblum D, Trochim W: A virtual national laboratory for reengineering clinical translational science. Sci Transl Med 2012;4:1–4.
48 CTSA Evaluation Key Function Committee: Evaluation Guidelines for the Clinical and Translational Science Awards (CTSAs). National CTSA Consortium Evaluation Key Function Committee, 2012. www.ctsacentral.org.
49 Rubio DM, et al: Evaluation metrics for biostatistical and epidemiological collaborations. Statistics in Medicine, 2011. http://onlinelibrary.wiley.com/doi/10.1002/sim.4184/pdf.
50 Clinical and Translational Science Awards Consortium: Community Engagement Key Function Committee Task Force on Principles of Community Engagement: Principles of Community Engagement, ed 2. NIH Publication No. 11–7782. Bethesdsa, NIH, 2011.
51 Fealing KH, et al (eds): The Science of Science Policy: A Handbook. Palo Alto, Stanford University Press, 2011.
52 Falk-Krzesinski HJ, et al: Advancing the science of team science. Clin Translation Sci 2010;3:263–266.
53 Falk-Krzesinski HJ, et al: Mapping a research agenda for the science of team science. Res Eval 2011;20:143–156.

Cathleen Kane, PhD
Assistant Director of Evaluation
Weill Cornell Clinical Translational Science Center
407 E. 61st, Room RR20
New York, NY 10065 (USA)
E-Mail cmk42@cornell.edu

Alving B, Dai K, Chan SHH (eds): Translational Medicine – What, Why and How: An International Perspective.
Transl Res Biomed. Basel, Karger, 2013, vol 3, pp 120–122 (DOI: 10.1159/000343464)

Envisioning the Future of Translational Medicine: A Snapshot of 2016

Barbara Alving[a] · Kerong Dai[b] · Samuel H.H. Chan[c]

[a]Uniformed Services, University of the Health Sciences, Bethesda, Md., USA; [b]Clinical Translational Center of Stem Cell and Regenerative Medicine, Shanghai Jiao Tong University School of Medicine, Shanghai, PR China; [c]Center for Translational Research in Biomedical Sciences, Chang Gung Medical Center, Kaohsiung, Taiwan

Numerous academic health centers (AHCs) throughout the world are beginning to embrace and fully develop the philosophy of translational medicine. Their focus is on training, building public-private partnerships, enhancing clinical research quality and efficiency, and developing community engagement. They all have the common goal of providing to patients and communities improved prevention and treatment strategies for diseases in an efficient, cost-effective, high-quality manner. If one arbitrarily defines 'the future' as 2016, what can we expect to find? Snapshots of progress can be captured in three general areas: leadership at AHCs, training programs for investigators, and engagement with communities.

In 2016, the leadership at AHCs will have fully established opportunities for investigators to easily access the core resources that are so essential to their laboratory and clinical work, as well as access to teams of experts who can provide consultation during the development of new studies. Small amounts of pilot funds will be used routinely to generate research projects that can be the foundation for obtaining grants for more comprehensive research. Interdisciplinary research studies will be developed and launched with appropriate review in a highly efficient fashion, due to utilization of Web-based procedures and the minimization of the number of steps required to initiate the study.

Continuous process improvement in the ways in which research studies are developed and monitored will be in place, allowing efficiencies in the conduct of clinical research. Monitoring and evaluation of the studies once they are begun will also be conducted to determine if any special assistance is required. AHCs will have established robust technology transfer offices that actively scan the university laboratories for novel ideas and provide short courses and updates on subjects such as intellectual property. The AHCs will have streamlined and standardized processes, such as memoranda of understanding with private enterprises, that were once unnecessarily labor and time intensive. Conflicts of interest will be managed in a transparent fashion, according to codes of conduct that are universally accepted.

There will continue to be the infrastructure for translational medicine in a generic fashion. However, institutions that focus on particular specialties will be applying many of the principles of translational medicine to their own research centers and laboratories, as well as using the generic infrastructure that has been put in place. In China, the more than 50 translational research centers that have been developed and modeled on the strength of the specialties in those centers will have had a chance to mature and create new ways of conducting translational medicine. Some translational research centers, such as those in Beijing, Shanghai, and Shenzhen, will have greater advantages in size, scale, and technology.

Leadership in AHCs throughout the world will be using better management principles and tools with respect to strategic thinking, project management, and interactions with the private sector. They will have developed even closer ties with business schools to ensure that training in these areas is available for investigators at all stages of academic development. AHCs will also find ways in which to reward and promote team science among their investigators. AHCs will continue to receive frank input and advice from their peers, as well as those in other fields. One way in which translational medicine centers will be receiving such advice is through external advisory boards convened several times each year to review the efforts of the individual centers. These boards may be comprised of investigators from other centers and also persons from biotech fields, as well as those who are philanthropists and/or venture capitalists. Ideas and processes will be shared internationally and then adopted as appropriate.

For investigators, training in translational medicine may be a part of their specialty training or be initiated at an earlier time in their careers since they will be able to take specific courses or enroll in degree-granting programs. By 2016, many such investigators will have been trained, and they will be comfortable convening teams to develop studies and protocols. They will be able to establish multisite clinical studies and trials more quickly because institutional review board (IRB) reciprocity will be in place at the AHCs, in which one IRB from one of multiple sites involved in the study will assume responsibility for reviewing and approving the protocol. The severe time constraints for those clinical investigators who are seeing patients will remain. However, PhDs will have been trained in the conduct of clinical trials and will be able to lead or be a crucial part of clinical research teams. The concept and familiarity with having a project manager as part of the team will have been fully accepted, and investigators will carefully review the most appropriate utilization of the talents on their teams in designing and initiating studies. Web-based technology will play a critical role in easing the challenges of clinical research, from designing a study and receiving approval from the IRB, to recruitment and enrollment of participants, to data entry.

AHCs will be active in their communities, establishing a sustained presence, listening to the community needs, and involving community leaders in decisions concerning research studies that impact the community. Realizing that effective community engagement requires the knowledge and experience of those in the social sciences and behavioral sciences, investigators will form interdisciplinary teams for effective community engagement research. Partnerships between AHCs and many other agencies, such as governmental (city, state), nonprofit, and commercial, will be developed as these agencies see a sustained engagement of AHCs with the community. With infrastructure in place to engage and work with community members, investigators will be conducting research studies in specific areas of health that are important to community members, such as mental health, heart disease, hypertension, diabetes, and cancer.

The current focus on translational research at a national level means that for each country engaged in the work, infrastructure is being developed that will ease the tortuous path from the basic laboratory to the community. Funding to support translational medicine initiatives will continue to be key to their progress and sustainability. In the USA, the National Institutes of Health will remain the key funder of such efforts. However, these funds will be increasingly leveraged by the AHCs and will be supplemented by state and private funds, including those from philanthropists. The experience in Taiwan for a private industry to support the translational effort may set an example for other companies in Taiwan and beyond in this endeavor. In China, the government will continue its commitment to funding translational research at numerous institutions; such research infrastructure will also be supplemented by funds from the provinces and city governments. Meanwhile, the national government will continue to emphasize the overall planning, ensuring that the sources and distribution of funding are more standardized and reasonable. The variety of funding sources will be an indication of strong and continued endorsement by multiple stakeholders.

Those who have had training in translational medicine will be well equipped to provide great value to policy making in health care delivery. By 2016, they will be leaders in organizations in the USA, such as the Centers for Medicare and Medicaid Services, a large federal funder of health services. In the UK, the policy leaders of the national health care system will be utilizing the advances made in research and care delivery to further guide their strategic plans for optimizing patient care at the bedside and in the community. The chronic, noncommunicable diseases that populations around the world are confronting are similar among developed and developing nations. However, each country is unique with respect to culture, patient expectations, history, geography, and size and degree of heterogeneity of the population. Nonetheless, translational medicine will be providing the infrastructure that is remarkably similar among countries, a platform and a philosophy on which solutions to the major universal challenges in human health can be more fully addressed on a collaborative national and international scale.

Author Index

Subject Index